The Catholic Biblical Quarterly
Monograph Series
36

Stockmen from Tekoa, Sycomores from Sheba

A Study of Amos' Occupations

BY

Richard C. Steiner

The Catholic Biblical Quarterly
Monograph Series
36

Produced in the United States of America

Library of Congress Cataloging-in-Publication Data

Steiner, Richard, 1945-
Stockmen from Tekoa, sycomores from Sheba : a study of Amos' occupations / by Richard C. Steiner. — 1st ed.
p. cm. — (The Catholic Biblical quarterly. Monograph series ; 36)
Includes bibliographical references and index.
ISBN 0-915170-35-3
1. Amos (Biblical prophet)—Career in herding. 2. Amos (Biblical prophet)—Career in agriculture. 3. Bible. O.T. Amos I, 1—Criticism, interpretation, etc. 4. Bible. O.T. Amos VII, 14-15—Criticism, interpretation, etc. 5. Balas (The Hebrew word) 6. Herding—Palestine. 7. Agriculture—Palestine. I. Title. II. Series.
BS580.A6S64 2003
224'.8067 dc22

2003019378

לזכר נשמת חותני מורי

מו"ה יששכר דוב בן מו"ה נתן נטע רויזנשיין

תנצב"ה

Contents

Acknowledgments • *ix*

Introduction • *1*

1. בולס שקמים: History of Interpretation • *5*
   Ancient and Medieval Interpretations of בולס • *5*
   The Septuagint's Interpretation of בולס and Sycomore Horticulture • *8*
   The Septuagint's Translation of שקמים: The Meaning and Etymology of συκάμινος • *17*
   Aquila's Interpretation of בולס • *20*
   Rashi's Interpretation of בולס and Sycomore Silviculture • *23*

2. בולס: Etymology and Meaning • *32*
   Bochart's Etymology of בולס • *32*
   Bochart's Etymology and a Post-biblical Survival of בלס "Sycomore Fig" • *35*
   Bochart's Etymology and the Meaning of בולס שקמים • *43*

3. בלס and שקמה: Linguistic Evidence for the Origin of the Biblical Sycomore • *48*
   The Controversy Surrounding the Origin of the Biblical Sycomore • *48*

The Distribution of *Bls* and *Šqmt* in the Semitic Languages • *52*
בלס and שקמה: Lexical and Botanical Imports from South Arabia • *58*
The Etymology of שקמה • *64*

4. בוקר, בנקדים, AND מאחרי הצאן • *66*
The Meaning of בוקר • *66*
The Meaning of בנקדים • *70*
The Syntax of דברי עמוס אשר היה בנקדים מתקוע • *87*
The Meaning of ויקחני ה׳ מאחרי הצאן • *91*

5. AMOS' OCCUPATIONS • *95*
The Herdsmen from Tekoa • *95*
The Location of Amos' Sycomores • *101*
Linking the Two Occupations • *105*
Amos' Occupations and His Prophecies • *116*

SUMMARY • *120*

BIBLIOGRAPHY • *123*
Ancient and Medieval Works • *123*
Modern Works • *127*

INDEX • *149*

Acknowledgments

This book grew out of a graduate course given at Yeshiva University, and so I would like to begin by thanking members of its faculty, student body and staff. Arthur Hyman, who became dean of the Bernard Revel Graduate School at a time of great turmoil, has worked hard to restore a sense of tranquility and an atmosphere in which scholarship is prized. David Berger, Louis H. Feldman, Elazar Hurvitz, S. Z. Leiman and Hayim Tawil, my friends, teachers and colleagues at Yeshiva University, have contributed to this book in numerous ways, as have my students A. Koller and M. T. Novick. Pearl Berger, dean of libraries, has greatly increased the resources available for research. Zvi Erenyi, Mary Ann Linahan, Haya Gordin, Kerry Santoro and other librarians of Yeshiva University have responded to my unending stream of requests with an alacrity that went well beyond the call of duty.

I would also like to express my deepest appreciation to the experts that I turned to outside of my institution: Joshua Blau and John Huehnergard in Semitics; John Brinkman and Erica Reiner in Assyriology; Leo Depuydt, Janet H. Johnson and Robert K. Ritner in Egyptology; Jan Joosten in ancient biblical translations; Harry Fox, Shama Friedman, B. Septimus, and Israel M. Ta-shma in Jewish Studies; Avinoam Danin, David Heller and Mordechai Kislev in botany. Their individual contributions are mentioned in many footnotes, but I am keenly aware that this does not give a full accounting of the generous and invaluable help I have received from them. The same goes for Mark S. Smith, who honored this work with a place in the CBQ monograph series and labored patiently and selflessly to improve every aspect of it.

Thanks are also due to my daughters and sons-in-law for their forbearance and assistance. Menachem (Manuel) Jacobowitz checked my treatment of rabbinic sources and provided references; Owen Cyrulnik made himself available at all hours to solve every conceivable kind of computer problem; and Joseph Crystal helped with a question in chemistry. Words cannot do justice to the debt that I owe to my wife, Sara. Suffice it to say that without her constant loving support, this book could not have been written. To all those who learn anything of value from this book, I say, in the words of Rabbi Akiva, שלי ושלכם שלה הוא.

תושלב״ע

New York
September, 2002

Richard C. Steiner

Introduction

Amos is viewed by many as a pivotal figure in the history of prophetism, in part because he worked for a living before his call instead of training to become a prophet. But what precisely was his occupation? The question has been discussed time and again in biblical scholarship, for "it is believed that as his sociocultural background is grasped or reconstructed we have a key, if not *the* key, to his message."[1]

The question ought to be easy to answer. After all, three verses of the book of Amos deal with the question. Biblical scholars are accustomed to making do with far less! In truth, the question "seems quite complex."[2] According to a scholar who devoted a good part of his career to just one of Amos' occupations, the answer requires input from the fields of botany, zoology, genetics, philology, Talmud, and Egyptology, not to mention Bible![3] Just reading the literature on the subject is a daunting task.

Two of the three verses treated in this monograph are contained in Amos' famous retort to Amaziah (Amos 7:14-15): לא נביא אנכי ולא בן

[1] G. F. Hasel, *Understanding the Book of Amos* (Grand Rapids, Mich., 1991) 29. Hasel himself devotes twelve pages to the question.

[2] J. A. Soggin, *The Prophet Amos* (London: SCM, 1987) 10. Soggin (*Amos*, 10-11) presents four distinct answers.

[3] J. Galil, השקמה בתרבות ישראל, *Teva Vaaretz* 8 (1966) 354 fig. 5.

נביא אנכי כי בוקר אנכי ובולס שקמים. ויקחני ה׳ מאחרי הצאן ויאמר אלי ה׳ לך הנבא אל עמי ישׂראל "I am not a prophet and I am not the son of a prophet; I am a בוקר and a בולס of sycomores. But the Lord took me away from אחרי הצאן and the Lord said to me, 'Go prophesy to my people Israel.'" The third verse is the book's superscription (1:1): דברי עמוס אשר היה בנקדים מתקוע "the words of Amos, who was among the נקדים from Tekoa."[4] In these three verses, we have no fewer than four expressions referring to Amos' occupations. Before he became a prophet, Amos was בנקדים and אחרי הצאן; he was a בוקר and a בולס שקמים.

At first glance, we seem to be faced with an embarrassment of riches. However, there is less usable information here than meets the eye, for, as one scholar laments: "Unique or near unique expressions dog our quest of the historical Amos!"[5] This is true of the three participles found in the four expressions: נוקד, בוקר, and בולס. Morphologically, these participles are *qal*, but lexicographically they are difficult, for they are poorly attested in Hebrew. The first occurs one other time in the Bible, enough to provide a general picture of its meaning but no more. The second occurs only here, but is arguably related either to the noun בקר "cattle" or to the *piel* verb בקר "examine." The third is the most isolated of all: even its root is unattested elsewhere in the Bible. As for the phrase אחרי הצאן, it too is rare in the Bible.

It is not surprising, therefore, that the meaning of each of these terms is mired in controversy. Moreover, the basic, lexicographic controversies have spilled over into other areas, spawning subsidiary controversies. Many scholars believe that 7:14, as traditionally interpreted, contradicts both 7:15 and 1:1; others deny this. Among the former group, many believe that the alleged contradictions call for the tools of higher or lower criticism, while others believe that the tools of philology suffice. Some scholars believe that Amos had two occupations; others, only one. Among the latter group, some believe that Amos was a herdsman, while others view him as a fruit farmer. Additional controversies arising from connotations of the aforementioned terms concern Amos' socioeconomic and sacral status.

[4] For a comprehensive *Forschungsgeschichte*, see the commentary and notes of M. Weiss, ספר עמוס (Jerusalem: Magnes, 1992) to these verses. My indebtedness to Weiss' painstaking labors will be evident throughout this monograph.

[5] A. G. Auld, *Amos* (Sheffield: JSOT Press, 1986) 39.

In this monograph I shall address these problems and others. I shall delve deeply into the practice of sycomore horticulture, sycomore silviculture and animal husbandry. I shall attempt to resolve the lexicographic controversies using the resources of Akkadian, Mishnaic Hebrew, Yemeni Arabic, etc. The resolution of these controversies will prove to have significance beyond the confines of Hebrew philology, biblical criticism, the history of prophetism, and the history of agriculture. Specifically, I shall attempt to show that our results shed light on the origin of the biblical sycomore—a problem that has generated much controversy among archeobotanists and paleobotanists—and the Arabian trade that began in the Late Bronze Age. In view of the complexity of the issues, the reader who does not prefer to be kept in suspense may wish to read the summary of conclusions at the end before becoming immersed in the details.

Finally, a word about my use of the spelling "sycomore" instead of the more common "sycamore." Both forms are used popularly to designate *Ficus sycomorus* L.,[6] a relative of the common fig tree also known as "sycamore/sycomore fig," "mulberry fig," "fig-mulberry," "Egyptian fig," "pharaoh's fig," "wild fig," "cluster fig," and "ass fig (Eselsfeige)." Unfortunately, the name "sycamore" is applied to unrelated trees in America and England. To avoid confusion, some botanists in recent years have recommended using the spelling "sycomore" for *Ficus sycomorus* and reserving "sycamore" for *Platanus*

[6] For other names found in older botanical works, see H. Grafen zu Solms-Laubach, *Die Herkunft, Domestication und Verbreitung des gewöhnlichen Feigenbaums (Ficus Carica L.)* (Abhandlungen der Königlichen Gesellschaft der Wissenschaften zu Göttingen 28; Göttingen: Dieterich, 1882) 103-6; and C. C. Berg and J. T. Wiebes, *African Fig Trees and Fig Wasps* (Amsterdam: North-Holland, 1992) 74. The terms *Ficus gnaphalocarpa* (Miq.) Steud. ex A. Rich. and (less commonly) *Ficus trachyphylla* Fenzl. are still used as synonyms of *Ficus sycomorus* L. in studies of African plants; see H.-J. von Maydell, *Trees and Shrubs of the Sahel* (Eschborn: Deutsche Gesellschaft für Technische Zusammenarbeit, 1986) 273; and Food and Agriculture Organization of the United Nations, *Traditional Food Plants: A Resource Book for Promoting the Exploitation and Consumption of Food Plants in Arid, Semi-arid, and Sub-humid Lands of Eastern Africa* (Rome: Food and Agriculture Organization of the United Nations, 1988) 288. Two subspecies of *Ficus sycomorus* have been distinguished—*Ficus sycomorus sycomorus* (the "common cluster fig") and *Ficus sycomorus gnaphalocarpa* (the "false cluster fig," in Namibia, Angola and West Africa)—but Berg and Wiebes (*African Fig*, 76) argue that the distinction is not sharp, "making recognition of subspecies hardly worthwhile."

occidentalis in America and *Acer pseudoplatanus* in England.[7] I have followed their recommendation in this monograph.

[7] See N. Hareuveni, *Tree and Shrub in Our Biblical Heritage* (trans. H. Frenkley; Kiriat Ono, Israel: Neot Kedumim, 1984) 41; F. N. Hepper, *Illustrated Encyclopedia of Bible Plants* (Leicester: Inter-Varsity Press, 1992) 114; idem, *Pharaoh's Flowers* (London: HMSO, 1990) 59; R. Gale, et al., "Wood," in *Ancient Egyptian Materials and Technology* (P. T. Nicholson and I. Shaw, eds.; Cambridge: Cambridge University Press, 2000) 340. Cf. also M. G. Easton, *Illustrated Bible Dictionary* (Grand Rapids, Mich.: Baker Book House, 1978) 668: "**Syc'amore**, more properly sycomore."

CHAPTER I

בולס שקמים: History of Interpretation

Ancient and Medieval Interpretations of בולס

An astonishing variety of interpretations of בולס can be found in ancient and medieval exegetical literature: "scratch, cut open (fruit of)" (LXX, Jerome, Theodoret of Cyrrhus, R. Hai Gaon *apud* Ibn Balʿam, Levi b. Yefet, Yeshuʿah b. Yehudah, Ibn Ezra);[1] "search" (Aquila, Rashi, Qara, Joseph Qimḥi);[2] "have" (Symmachus, Targum Jonathan);[3] "pick, gather (fruit of)" (Peshiṭta, David Qimḥi, Abarbanel);[4]

[1] Origen, *Origenis Hexaplorum* (ed. F. Field; Oxford: Clarendon Press, 1875) 2. 978 n. 17; Jerome, *Commentarii in Prophetas Minores* (Turnholti: Brepols, 1969-70) 1. 324 lines 402-4; S. A. Poznański, "The Arabic Commentary of Abu Zakariya Yaḥya (Judah ben Samuel) Ibn Balʿam on the Twelve Minor Prophets," *JQR* 15 (1924-25) 32; David ben Abraham al-Fāsī, *Kitāb Jāmiʿ al-Alfāẓ of David ben Abraham al-Fāsī* (ed. S. L. Skoss; New Haven: Yale University Press, 1936-45) 1. 203 (first apparatus); Ibn Ezra, שני פירושי ר׳ אברהם אבן עזרא לתרי־עשר (ed. U. Simon; Ramat-Gan: Bar-Ilan University, 1989) 311. For Yeshuʿah b. Yehudah, see below.

[2] Origen, *Hexapla*, 2. 978; Joseph Qimḥi, ספר הגלוי (ed. H. J. Mathews; Berlin: Mekize Nirdamim, 1887) 75.

[3] Origen, *Hexapla*, 2. 978; *The Bible in Aramaic* (ed. A. Sperber; Leiden: E. J. Brill, 1959-73) 3. 427.

[4] *The Old Testament in Syriac According to the Peshiṭta Version* (Leiden: E. J. Brill, 1972-) 3/4. 33; Isaac Abarbanel, *Don Isaac Abrabanel y su comentario al Libro de Amos* (ed. G. Ruiz; Madrid: UPCM, 1984) 196.

"tend" (Al-Qumisi, Isaiah of Trani);[5] "carry, transport (fruit of)" (Yefet);[6] "cut off" (Al-Fāsī);[7] "mix (leaves of)" (R. Sherira Gaon *apud* Ibn Janāḥ, Ibn Parḥon, Tanḥum Yerushalmi);[8] "beat (leaves off of)" (Ibn Janāḥ, Ibn Parḥon);[9] "shake (leaves off of)" (Menaḥem, Tanḥum Yerushalmi);[10] "dry" (Ibn Ezra);[11] and "pasture in the shade of" (Eliezer of Beaugency).[12] A few of the exegetes (e.g., Jerome, R. Hai Gaon, Rashi) buttress their interpretations with snippets of information about the cultivation of the sycomore in ancient and medieval times.

In the modern period, as interest in *realia* grew, biblical scholars began to delve more deeply into the ethnobotanical context of the problem. S. Bochart was a pioneer in this regard; his brilliant discussion of the etymology of בולס in *Hierozoicon*, published in 1663,[13] includes descriptions of the sycomore excerpted from classical authors. In 1782-83, the rich literature on the natural history of the sycomore was summarized for biblical scholars by H. E. Warnekros in a seventy-page article.[14] More recently, much light has been shed on

[5] Daniel Al-Qumisi, פתרון שנים עשר (ed. I. D. Markon; Jerusalem: Mekize Nirdamim, 1957) 38; Isaiah of Trani, פירוש נביאים וכתובים (ed. A. J. Wertheimer; 2nd ed.; Jerusalem: Ktab Yad Wasepher, 1978) 2. 98.

[6] Ms. British Library Or. 2400 (Margoliouth 282), p. רב = f. 102b, l. 5: חאמל פלק אלגמיז "a transporter of sycomore fig halves"; l. 13 וכנת אחמל פלק אלגמיז ואביעהא "I used to transport sycomore fig halves and sell them." The former is from the translation of the verse; the latter is from the commentary.

[7] Al-Fāsī, *Jāmiʿ al-Alfāẓ* 1. 203 line 3.

[8] Jonah Ibn Janāḥ, *Kitāb al-ʾuṣūl: The Book of Hebrew Roots* (Oxford: Clarendon Press, 1875) 96 lines 5-9; Solomon b. Abraham Ibn Parḥon, מחברת הערוך (ed. S. G. Stern; Pressburg: Typis Antonii Nobilis de Schmid, 1844) 2. 9b; Tanḥum Yerushalmi, פירוש תנחום בן יוסף הירושלמי לתרי־עשר (ed. H. Shy; Jerusalem: Magnes, 1991) 94-95.

[9] Ibn Janāḥ, *ʾUṣūl*, 96 lines 1-4; Ibn Parḥon, מחברת הערוך, 2. 9b.

[10] Menaḥem b. Saruq, *Maḥberet* (ed. A. Sáenz-Badillos; Granada: Universidad de Granada, 1986), 85*; Tanḥum Yerushalmi, פירוש לתרי־עשר, 94-95.

[11] Abraham Ibn Ezra, פירושי תרי־עשר (ed. U. Simon; Ramat-Gan: Bar-Ilan University, 1989-) 249-50.

[12] Eliezer of Beaugency, *Kommentar zu Ezechiel und den XII kleinen Propheten* (ed. S. A. Poznański; Warsaw: Mekize Nirdamim, 1909-13) 2. 152. For an annotated Spanish translation of this and other medieval Jewish commentaries on Amos, see *Comentarios hebreos medievales al libro de Amos* (ed. G. Ruiz González; Madrid: UPCM, 1987).

[13] See below.

[14] H. E. Warnekros, "Historia naturalis Sycomori ex veterum botanicorum Monumentis et Itinerariis conscripta," *Repertorium für Biblische und Morgenländische Litteratur* 11 (1782) 224-71, 12 (1783) 81-104.

the problem by students of the sycomore from outside of the field of biblical studies, in particular, E. Sickenberger, C. Henslow, L. Keimer, T. W. Brown and F. G. Walsingham, and J. Galil.[15] Galil studied the tree in Israel, Cyprus and Kenya, and devoted a series of articles and part of a monograph to it over a period of two decades. An even longer love-affair with the sycomore was carried on by G. Schweinfurth; it is said that he spent fifty-one years studying the sycomore in Palestine, Egypt, Ethiopia and Yemen.[16] Two other botanists—Goldmann and Löw—compiled exhaustive surveys of the literature.[17] Much of this research has been made available to biblical scholars in an article by T. J. Wright.[18]

Despite all of the work that has been done, a number of scholars feel that the problem has not been resolved in a satisfactory manner. After reviewing the evidence, J. H. Hayes writes: "In spite of such evidence, we still do not know what function a *boles* sycomores performed."[19] G. V. Smith feels that "none of these alternate translations are overly convincing."[20] F. E. Greenspahn writes: ". . . one must conclude that the evidence is not sufficient to permit a convincing treatment of this word."[21] S. N. Rosenbaum suggests that the generally accepted etymology of בולס "takes us in the wrong direction."[22] M. Weiss concludes that "the etymology of בולס is problematic."[23] A closer look at the history of this word before and after Amos is needed to dispel these nagging doubts.

This is not the place to examine in detail all of the many interpretations listed above. This chapter explores three of the most interesting

[15] See below.

[16] See Keimer, "Eine Bemerkung zu Amos 7,14," *Bib* 8 (1927) 442.

[17] F. Goldmann, *La figue en Palestine à l'époque de la mischna* (Paris: Librairie Durlacher, 1911) 42-46; I. Löw, *Die Flora der Juden* (Vienna and Leipzig: R. Lowit, 1924-34) 1. 274-80.

[18] T. J. Wright, "Amos and the 'Sycomore Fig,'" *VT* 26 (1976) 362-68.

[19] J. H. Hayes, *Amos, the Eighth-Century Prophet* (Nashville: Abingdon, 1988) 238.

[20] G. V. Smith, *Amos: A Commentary* (Grand Rapids, Mich.: Zondervan, 1989) 240.

[21] F. E. Greenspahn, *Hapax Legomena in Biblical Hebrew: A Study of the Phenomenon and Its Treatment Since Antiquity with Special Reference to Verbal Forms* (SBLDS 74; Chico, Calif.: Scholars Press, 1984) 106.

[22] S. N. Rosenbaum, *Amos of Israel: A New Interpretation* (Macon, Ga.: Mercer University Press, 1990) 48.

[23] Weiss, ספר עמוס, 1. 239.

ones. Although we shall ultimately reject them, they serve as an introduction to some of the points made below. We shall return to some of the others in later chapters.

The Septuagint's Interpretation of בולס and Sycomore Horticulture

The Septuagint's rendering of בולס שקמים is κνίζων συκάμινα. In my opinion, the best translation of this Greek phrase is "a scratcher of sycomore figs." This translation requires some justification. The verb κνίζω in this phrase has been translated in a great variety of ways, but "scratch" is the first meaning given for it by Liddell and Scott,[24] and we shall see below that it fits the context perfectly. The meaning and etymology of συκάμινα will be treated in the next section.

As Jerome and Theodoret of Cyrrhus knew, the Septuagint's rendering of בולס שקמים refers to the practice of lacerating the figs on the sycomore tree to make them ripen.[25] This practice is attested very early in Egypt, long before the time of Amos. We shall see below that it is also attested in Palestine, albeit at a later period.

The connection between lacerating and ripening has been the subject of speculation since ancient times. Theophrastus seems to have believed that the effluence of juice from the wounded "Cyprian figs" causes them to ripen.[26] Jerome writes that, without this operation, the figs are spoiled by "gnats" (=wasps?), while Bar Bahlul suggests that the practice induces ripening by causing "gnats" to enter.[27]

Until thirty-five years ago, modern students of the sycomore held similar views. Botanists connected the operation with the wasps that

[24] H. G. Liddell and R. Scott, *A Greek-English Lexicon* (Oxford: Clarendon Press, 1996) 965.

[25] See Origen, *Hexapla*, 2. 978 n. 17; Jerome, *Commentarii*, 324 lines 402-4. (For the interpretations of these and other church fathers, see U. Treu, "Amos VII 14, Schenute und der Physiologos," *NovT* 10 [1968] 234-40.) By contrast, Theodore of Mopsuestia (*Theodori Mopsuesteni Commentarius in XII Prophetas* [Wiesbaden: O. Harrassowitz, 1977] 147 lines 2-3) took κνίζων to mean περισκάλλων "hoeing around (?)" and γεωργῶν "plowing (?)".

[26] Theophrastus, *Enquiry into Plants* (trans. A. Hort; Cambridge, Mass.: Harvard University Press, 1948) 1. 292-95 (4.2.3). Cf. Wright, "Sycomore Fig," 364.

[27] Jerome, *Commentarii*, 324 lines 401-4; Hasan Bar Bahlul, *Lexicon Syriacum auctore Hassano Bar-Bahlule* (ed. R. Duval; Paris: e Reipublicae typographaeo, 1888-1901) 1115 lines 8-9.

lay their eggs in the female flowers within the figs:[28] either the gashing served to let the wasps out or it served to let the air in, thereby drying up the female flowers upon which the wasps depend.[29] One biblical scholar elaborated on Theophrastus' theory:

> . . . the fruit is rather bitter, but by making an incision in them before they are ripe, one can make some of the juice run out. The rest then ferments, and gives the fruit a sweet taste. . . .[30]

Galil's experiments showed the true reason: gashing stimulates the production of ethylene, a gas that is used commercially for the ripening of oranges, bananas, etc. With the sycomore fig, it acts as a growth stimulator as well as a ripener, inducing a very great increase in size and weight.[31]

At first glance, the Septuagint's use of the verb κνίζω "scratch" to refer to this practice seems a bit odd. All of the ancient and modern evidence indicates that the laceration in question was more of a gash than a scratch. The solution to this problem lies in the ancient Greek accounts of this practice. It has often been noted that the same Greek verb or a variant of it is used in descriptions of the sycomore in Egypt.[32] The earliest such description is that of Theophrastus:

[28] The (mutualistic and parasitic) relationships of (pollinating and non-pollinating) fig wasps to their figs are currently the subject of intensive research, stimulated in part by Galil's studies with D. Eisikowitch and others. For extensive bibliography, see Berg and Wiebes, *African Fig* 273-85; G. D. Weiblen, *Phylogeny and Ecology of Dioecious Fig Pollination* (Ph. D. diss., Harvard University, 1999) 261-304; C. Kerdelhué, et al., "Comparative Community Ecology Studies on Old World Figs and Fig Wasps," *Ecology* 81 (2000) 2847-49; and add J. Galil and D. Eisikowitch, "Further Studies on the Pollination Ecology of *Ficus Sycomorus* L. (Hymenoptera, Chalcidoidea, Agaonidae)," *Tijdschrift voor Entomologie* 112 (1969) 1-13.

[29] The latter explanation is still given by M. N. el-Hadidi and L. Boulos, *The Street Trees of Egypt* (Cairo: The American University in Cairo Press, 1988) 70.

[30] E. Hammershaimb, *The Book of Amos* (Oxford: B. Blackwell, 1970) 118. So too J. A. Arieti, "The Vocabulary of Septuagint Amos," *JBL* 93 (1974) 343.

[31] J. Galil, "An Ancient Technique for Ripening Sycomore Fruit in East-Mediterranean Countries," *Economic Botany* 22 (1968) 188-89.

[32] G. Henslow, "Egyptian Figs," *Nature* 47 (1892) 102; idem, "The Sycomore Fig," *Journal of the Royal Horticultural Society* 27 (1902) 130-1; S. R. Driver, *The Books of Joel and Amos* (Cambridge: Cambridge University Press, 1897) 212 n. 2; Wright, "Sycomore Fig," 363. Indeed, the point is implicit already in S. Bochart, *Hierozoicon*

καὶ πέττειν οὐ δύναται μὴ *ἐπικνισθέντα*· ἀλλ' ἔχοντες ὄνυχας σιδηροῦς *ἐπικνίζουσιν*· ἃ δ' ἂν *ἐπικνισθῇ* τεταρταῖα πέττεται·

It cannot ripen unless it is scratched on the surface; but they scratch it on the surface with iron claws; the fruits thus scratched ripen in four days.[33]

The word used by Theophrastus to describe the Egyptian practice is ἐπικνίζω "scratch on the surface."[34] Dioscorides uses that word as well, while Athenaeus uses κνίζω, the same verb as LXX.[35] We seem to be dealing with a technical term associated with sycomore horticulture in Egypt.

The reason for the use of verbs meaning "scratch" and "scratch on the surface" is clear from the context: the operation was performed with an instrument called a "claw." Indeed, according to the 1996 revised supplement to Liddell and Scott, several occurrences of κνίζω that were formerly thought to mean "tickle" actually mean "scrape or scratch with the fingernails."[36] It appears that κνίζω and ἐπικνίζω are used to refer to laceration with fingernails and claws.

Thanks to the work of Figari, Sickenberger, Henslow and Keimer, it is reasonably clear that the term "claw" used by Theophrastus refers to a knife or razor with a blade bent into the shape of a hook or claw. The existence of such an instrument in ancient Egypt would explain the curved shape of the scars on many ancient Egyptian representations of sycomore figs.[37] It would also explain why both the hieratic bird's claw determinative and the knife determinative are used with the Egyptian word for the gashed sycomore fig.[38] Indeed, such an instru-

(London: J. Martyn & J. Allestry, 1663) 1. 383-84 (= S. Bochart, *Hierozoicon* [ed. E. F. C. Rosenmüller; Leipzig: Weidmann, 1793-96] 1. 405-6).

[33] Theophrastus, *Enquiry* 1. 292-93 (4.2.1).

[34] Liddell and Scott, *Lexicon*, 639.

[35] Pedanii Dioscuridis, *De Materia Medica* (ed. M. Wellmann; Berlin: Weidmann, 1906-14) 1. 116 lines 14-15; Athenaeus, *The Deipnosophists* (trans. C. B. Gulick; Cambridge, Mass.: Harvard University Press, 1927-41) 1. 222 (2.51).

[36] Liddell and Scott, *Lexicon*, Supplement, 179.

[37] L. Keimer, "Sur quelques petits fruits en faïence émaillée datant du Moyen Empire," *Bulletin de l'Institut français d'archéologie orientale* 28 (1929) 52-55.

[38] L. Keimer, "An Ancient Egyptian Knife in Modern Egypt," *Ancient Egypt* 1928, 65; idem, "Sprachliches und Sachliches zu ελκω 'Frucht der Sykomore,'" *AcOr* 6 (1928) 293-94; idem, "Petits fruits," 55. Egyptian has two words for the sycomore fig, one of which is believed to denote the gashed ("notched") fruit; see chapter 3 below.

ment was still being used to gash sycomore figs in some parts of Egypt in the nineteenth century and the beginning of the twentieth.[39]

The above explanation is strengthened by another detail in Theophrastus' account. In describing the practice of gashing another variety of sycomore fig, the "Cyprian fig" in Crete, he uses a different term: ἐπιτέμνω "cut on the surface."[40] The difference in terminology seems to reflect a difference in the instruments used. In Cyprus, only ordinary kitchen knives are used today to gouge the fruit.[41] If that was also the case in ancient times, it would help to explain why Theophrastus uses a verb meaning "cut on the surface" in discussing the "Cyprian fig" instead of the technical term meaning "scratch on the surface" that he uses in discussing the "Egyptian sycamine."

The LXX's interpretation of בולס reappears in Semitic sources in the Middle Ages. An Aramaic version of it seems to be attested in the tenth-century Syriac dictionary of Bar Bahlul. There we find the phrase מסטף שקמ̈א—presumably a rendering of בולס שקמים—defined as "prick with a needle so that the gnats enter and it ripens."[42] Löw wonders about Bar Bahlul's source:

> Woher BB. die Wendung nimmt weiß ich nicht. Die Bibelübersetzer zu Amos 7, 14 kennen sie nicht. Hex. z.B. hat: חרט = κνίζων.[43]

I would like to suggest that Bar Bahlul took the phrase מסטף שקמ̈א from a version based largely on LXX, the Syropalestinian version, which is not extant on our verse.[44] As Löw himself notes, the verb סטף is attested in the fragments of that translation that *are* extant; it

[39] A. Figari (*Studii scientifici sull'Egitto e sue adiacenze compresa la penisola dell'Arabia Petrea* [Lucca: G. Giusti, 1865] 2. 178) describes "a kind of thimble (made) of sheet-tin ending in a claw." By the beginning of this century, the hooked (claw-shaped) blade had already evolved in the Cairo district and elsewhere into a circular (ring-shaped) blade, but Sickenberger (*apud* Henslow, "Sycomore Fig," 128-29) and Keimer ("Petits fruits," 54-55, 57-60) were able to find a few specimens with the older shape in Damietta and Miḥallet Marḥoum, respectively. See also Keimer, "Knife," 66; and T. W. Brown and F. G. Walsingham, "Sycamore Fig in Egypt," *The Journal of Heredity* 8 (1917) 10.

[40] Theophrastus, *Enquiry*, 1. 294-95 (4.2.3).

[41] Galil, "Ancient Technique," 186.

[42] Bar Bahlul, *Lexicon*, 1115 lines 8-9.

[43] Löw, *Flora*, 1. 280.

[44] See also n. 56 below.

renders κατατέμνω "lacerate" in Isa 15:2 (כול אידרעיא מסטפין).[45] Thus, my suggestion boils down to a claim that the Syropalestinian version uses the same Aramaic verb (סטף) to render κνίζω in Amos 7:14 that it uses to render κατατέμνω in Isa 15:2. Evidence for this claim can be adduced from the Syrohexapla, which uses the same Aramaic verb (חרט "scratch") to render κνίζω in Amos 7:14 (חָרט שקֿמא "a scratcher of sycomores") that it uses to κατατέμνω in Isa 15:2 (כלהון דרֿעא מחרֿטא "all arms scratched").[46]

The use of the verb סטף to render בולס שקמים is worthy of attention, for it appears to be the Aramaic and Mishnaic Hebrew technical term for the gashing of sycomore figs. Elsewhere in Bar Bahlul's dictionary, Syriac סטף is equated with Arabic *šaraṭa* "scarify, make an incision in the process of cupping."[47] The verb therefore refers to non-hostile (curative or self-inflicted) laceration.[48]

The Mishnah agrees with Bar Bahlul's citation in using this verb to refer to an operation performed on sycomore figs: כל בנות שקמה פטורות חוץ מן המסוטפות "all sycomore fruit is exempt [from tithing in cases of doubt] except for those that are מסוטפות" (*m. Dem.* 1.1). The term המסוטפות has often been taken as referring to sycomore figs that have split open naturally.[49] This interpretation is hardly likely for a verb in

[45] LXX to Isa 15:2 has πάντες βραχίονες κατατετμημένοι "all arms lacerated," equivalent to כל ידים גדודות, which is not in the Masoretic text there, but does appear in a parallel prophecy against Moab in Jer 48:37. The Greek word is also used to render the verb *ś-r-ṭ* in Lev 21:5 ובבשרם לא ישרטו שרטת = καὶ ἐπὶ τὰς σάρκας αὐτῶν οὐ κατατεμοῦσιν ἐντομίδας.

[46] *Codex Syro-hexaplaris Ambrosianus photolithographice editus* (ed. A. M. Ceriani; *Monumenta sacra et profana* 7; Milan: Impensis Bibliothecae Ambrosianae, 1874) 101a, 178a.

[47] See, for example, Bar Bahlul, *Lexicon*, 880 line 4. The verb *šaraṭa* is also used in the Arabic translation of Dioscorides in describing the laceration of sycomore fruit; see at n. 63 below.

[48] For more on this verb, see M. Moreshet, לקסיקון הפועַל שנתחדש בלשון התנאים (Ramat-Gan: Bar-Ilan University Press, 1980) 243. Cf. also A. Tal, התרגום השומרוני לתורה (Tel-Aviv: Tel-Aviv University, 1980-83) 1. 123, where במסטופיה renders ברהטים in Samaritan targ. A to Gen 30:38. It appears that the translator understood ברהטים as modifying פצל. In that case, the -ב is instrumental and a מסטוף is a tool. This suggestion is recorded in my name (with the translation "chisel") in A. Tal, *A Dictionary of Samaritan Aramaic* (Leiden: E. J. Brill, 2000) 2. 582.

[49] See Nathan b. Yeḥiel, ערוך השלם (ed. A. Kohut; Vienna: n.p., 1878-92) 6. 39a: שנבשלו באילן עד שנתבקעו מאליהן "that ripened on the tree until they split by them-

the *pual*, and it fails to elucidate the context. The true meaning of מסוטפות was given by R. Sherira Gaon in a responsum: "המסוטפות—gashed (משקקאת), the gashing of (or: a gash in) a wild fig is called סטף."[50] There can be no doubt about the meaning of משקקאת. The D-

selves." This definition is cited with minor variants in many later commentaries. Thus we find שנתבשלו באילן עד שנבקעו מאליהן in R. Samson of Sens, משניות מן סדר זרעים עם פירוש רבינו משה בר מיימון ופירוש רבינו שמשון משאנץ, 8b (printed near the end of vol. 1 of the standard Vilna edition of the Babylonian Talmud) and שנתבשלו באילן עד שנתבקעו in R. Solomon b. Joseph Sirillo, מסכת דמאי מן תלמוד ירושלמי . . . לר׳ שלמה בכ״ר יוסף סירילי׳או (ed. P. Shapiro and J. Freimann; Jerusalem: מסורה, 1955) 2b. See also Hayes, *Amos*, 238: "No ancient or modern evidence exists explicitly indicating that sycomore figs were slashed in Palestine . . . In the Mishnah, . . . sycomore figs are distinguished according to whether or not they burst open naturally on the tree (*m. Dem.* 1.1)." Z. Amar (גידולי ארץ־ישראל בימי־הבינים: תיאור ותמורות, [Ph.D. diss., Bar-Ilan University, 1996] 300-301) cites R. Solomon Sirillo as evidence that sycomore figs were still being gashed in 16th-century Palestine, but he is uncertain as to whether gashing is, in fact, what the Mishnah is talking about. This is precisely backwards. The Mishnah is certainly talking about gashing (see below), while Sirillo's definition is taken from the ערוך and refers to splitting as the result of ripening rather than vice versa.

[50] Contrast S. Assaf, פירוש ששה סדרי משנה לרבינו נתן אב הישיבה, *Kiryat Sefer* 10 (1934) 528 last two lines and Sherira Gaon, קטע מפירוש מלים של רב שרירא לסדרי טהרות וזרעים, in תשובות הגאונים מכתבי־יד שבגנזי קמברידג׳ (ed. S. Assaf; Jerusalem: Mekize Nirdamim, 1942) 179 lines 8-9. This definition has been consistently misunderstood, by both traditional scribes and modern scholars. It must have originally read: המסוטפות אלמשקקאת. אלשק מן/פי אלתוב יסמא סטף. The word תוב, "wild fig, caprifig," is rare in Arabic and Jewish Aramaic, but well attested in Syriac; see K. Brockelmann, *Lexicon Syriacum* (2nd ed.; Halis Saxonum: M. Niemeyer, 1928) 818a; I. Löw, *Aramäische Pflanzennamen* (Leipzig: W. Engelmann, 1881) 391; idem, *Flora*, 1. 233. R. Sherira uses it to gloss בנות שקמה in this mishnah (כל בנות שקמה פטורות חוץ מן המסוטפות), instead of the more usual expression for "wild fig," תין ברי, used by Maimonides in his commentary *ad loc* (see chapter 2 n. 29 below). The use of תוב to gloss שקמה and vice versa is known from the ninth-century commentary of Ishodad of Merv (*Commentaire d'Išoʿdad de Merv sur l'Ancien Testament* [ed. C. Van den Eynde; CSCO 303; Louvain: L. Durbecq, 1950-] 4. 90 line 9) to Amos 7:14 and other Syriac sources; see Bar Bahlul, *Lexicon*, 2005 line 17 and 2039 lines 10-11. Later Jewish scribes, who were not familiar with this rare term, mistook it for the common word ת֗וב "garment." This mistake is clearest in the manuscripts of the expanded Mishnah commentary of R. Nathan Av ha-Yeshivah, in which Sherira's definition is cited. Both MS British Library Or. 11117 and MS JTS R 1492 read המסוטפות נוע מן אלגמיז וקאל ר׳ שרירא אלמשקקאת לאן אלשק פי אלת֗וב יסמא סטף; the point over the ת is clearly visible in both. In T. S. Arab 18(1)[2], the sole surviving manuscript of Sherira's responsum (Sherira Gaon, קטע, 175, lines 8-9) the text reads: המסוטפות אלמשקקאת אלשק מן אלתוב איצא יסמא סטף. No point is visible over the ת (in the

stem passive form implies a human agent. Indeed, משקק in the *active* voice is used in Levi b. Yefet's compendium of Al-Fāsī's dictionary to gloss בולס in our verse.[51]

As Löw hints, this interpretation of מסוטפות makes perfect sense in the Mishnah. The difference between ungashed sycomore figs and gashed ones is that "jene gelten als herrenlos, diese als erwarteter Ertrag."[52] This brief comment apparently alludes to the view of R. Yoḥanan in *y. Dem.* 1.1, 21c, according to which the reason most sycomore fruit is exempt from tithing in cases of doubt is that it is presumed to have been abandoned. A man-made gash, unlike natural splitting open, shows that the fig was not abandoned.[53] The Mishnah shows conclusively that the gashing of sycomore fruit was known in

---

microfilm, at least), but that may be because the scribe regularly dispenses with the superior points, as he does in the following word, א׳צׄא "also." The insertion of the latter word seems to indicate that he too misunderstood the word תוב. Further confusion is evident in Maimonides' comment (משנה עם פירוש רבינו משה בן מימון, מקור ותרגום [ed. J. Qafiḥ; Jerusalem: Mossad Harav Kook, 1963-68] 1. 133): מסוטפות - משקקה̈ והו נוע מן אנואע ד̇לך אלתין "split, and that is one of the types of this fig." As it stands, this comment appears to be just a garbled version of R. Nathan's, in which the two alternative interpretations have been combined into one. It is possible that והו should be emended to או הו, but is also possible that Maimonides took משקקה̈ as referring to the variety of sycomore figs that requires gashing, as opposed to the variety that ripens without being gashed; see n. 54 below. In *Mishneh Torah*, *Hilkhot Maʿaser* 13,1, he mentions בנות שקמה but not בנות שקמה מסוטפות.

[51] Al-Fāsī, *Jāmiʿ al-Alfāẓ*, 1. 203 (first apparatus): ובולס שקמים וכ̇תאן אלג̇מיז וקיל משקק פלק אלג̇מיז "and a circumciser of sycomore figs. Others say: 'one who cuts sycomore figs in half.'"

[52] Löw, *Flora*, 1. 280.

[53] Cf. E. Sickenberger *apud* Henslow, "Egyptian Figs," 102: "The figs of the third generation are larger, of an agreeable taste, and sweet-scented; but they are not operated upon, only because in August and September, though the trees are much fuller of fruit than in May and June, the people have so much to do at that time. They are seldom sold, and only eaten by the owners of the trees, or else they are abandoned to the field-mice, birds and dogs. . . ." So too L. Reynier, "Méthode de caprification usitée sur le figuier sycomore," in *Mémoires sur l'Egypte publiés dans les campagnes du général Bonaparte* (Paris: P. Didot L'ainé, 1800-1803) 3. 189: "Ce procédé n'est plus usité dès que la seve (sic) commence à diminuer; alors on se borne à cueillir les fruits les mieux développés, et le reste tombe et pourrit au pied de l'arbre." See also Figari, *Studii*, 178: "The figs of the fall season ripen spontaneously without the intervention of cutting off their orifice. . . ."

ancient Palestine[54] and that it was denoted there by the verb סטף in the D-stem.[55] This strengthens our conjecture that the phrase מסטף שקְמא is from the Syropalestinian version.[56]

In Judeo-Arabic sources, the LXX's interpretation appears in the eleventh century. Ibn Balʿam cites the following comment in the name of R. Hai Gaon: "the cutting open of the sycomore (fruit) (תשריח אלגֻמיז), a well-known craft in Syria-Palestine."[57] The Karaites were

[54] Sycomore figs in modern Israel ripen without being gashed. According to Galil (השקמה, 349-52; "Ancient Technique," 188), the change is due to genetic mutation and natural selection. Amar (גידולי ארץ־ישראל, 300) believes that the variety of sycomore figs that requires gashing co-existed in medieval Palestine with the variety that ripens without being gashed.

[55] It has been claimed that the Mishnah has a second reference to the gashing of sycomore figs: סכין את הפגים ומנקבין אותן עד ראש השנה "they may oil the young figs and pierce them until the New Year" (m. Shebi. 2.5); see N. Hareuveni, *Tree and Shrub*, 91. However, this two-step procedure has survived until modern times, and almost all descriptions of it speak of the common (Carian) fig; see Maimonides' commentary to this mishnah; Thomas Shaw, 1722 and 1738, *apud* A. Goor, "The History of the Fig in the Holy Land from Ancient Times to the Present Day," *Economic Botany* 19 (1965) 134; Reynier, "Méthode," 186; Galil, השקמה, 348. Only one source—Daʾūd al-Anṭākī in sixteenth century Palestine—reports this procedure for both common figs and sycomore figs; see Amar, גידולי ארץ־ישראל, 303-4 (where, however, the word *ḥattā* "until" has been misread as *ḥayy*).

[56] According to M. Bar-Asher, it is natural to assume that Bar Bahlul had access to this work, but this is the first clear evidence for that assumption (personal communication).

[57] Poznański, "Ibn Balʿam," 32; Z. Amar, פירושים ריאליים בפרשנות ימי־הביניים לצומח של ארץ־ישראל, *Sinai* 116 (1995) 94 (I am indebted to D. Talshir for this reference). Despite the fact that R. Sherira Gaon seems to have been familiar with the practice of gashing sycomore figs (see above), he did not connect it with the phrase בולס שקמים, according to the testimony of Ibn Janāḥ, *ʾUṣūl*, 96 lines 5-9. It was left to his son, R. Hai Gaon, to make that connection. It is not impossible that he made it with the help of a Syriac rendering like Syrohexaplaric חָרט שקְמא "a scratcher of sycomores" or חָאך תוּבֿא "a scratcher of wild figs"; see at n. 46 above, and Ishodad of Merv *Commentaire*, 4. 90 line 9. Such a rendering could have been supplied to him by the same Nestorian catholicos (גֿאתליק) who supplied him with a Peshiṭta-like Syriac rendering of Ps 141:5; see the passage from Hai's biography cited in Joseph Ibn ʿAqnīn, התגלות הסודות והופעת המאורות: פירוש שיר השירים (ed. A. S. Halkin; Jerusalem: Mekize Nirdamim, 1964) 495. Whether that catholicos himself would have taken חָרט שקְמא to refer to the gashing of sycomore figs is another matter. Bar Ali uses חרט to gloss חרת "dig out, furrow," and Ishodad of Merv lists חָרט שקְמא as a variant of מחפר שקְמא "a digger of (the soil around) sycomores"; see R. Payne Smith, *Thesaurus Syriacus* (Oxford: Clarendon Press, 1879-1901) 1. 1370 s.v. חרט; *Commentaire d'Išoʿdad de Merv sur l'Ancien Testament* (trans. C. Van

particularly fond of this interpretation. One of the glosses of ובולס שקמים in Levi b. Yefet's compendium of Al-Fāsī's dictionary is וכתאן אלגמיז "and a circumciser of sycomore figs."[58] And in a recently published passage from Yeshuᶜah b. Yehudah's commentary to Exod 23:11, we have "[by] ואין מגמזין they mean לא יכתנון, one may not circumcise [sycomore figs during the sabbatical year], like ובולס שקמים."[59] Here, too, we are dealing with a technical term for the gashing of sycomore figs, this time in Arabic.[60]

The LXX's interpretation makes an appearance in the twelfth century, as well. According to Ibn Ezra, ובולס שקמים means ומשרטט השקמים שלי כדי שימתקו "and a scorer of my sycomores in order that they become sweet."[61] Although Ibn Ezra knew Ibn Balᶜam's work, his

---

den Eynde; CSCO 304; Louvain: L. Durbecq, 1950-) 4. 114 n. 4. It is possible that at least some Nestorians viewed חָרֵט שָׁקְמֵא and מחפר שָׁקְמֵא as being synonymous; cf. n. 25 above.

[58] See n. 51 above. The same rendering is found in a Coptic-Arabic dictionary; see n. 60 below.

[59] O. Tirosh-Becker, "Linguistic Study of a Rabbinic Quotation Embedded in a Karaite Commentary on Exodus," in *Studies in Mishnaic Hebrew* (ed. M. Bar-Asher; Jerusalem: Magnes, 1998) 385-88: ואין מגמזין ירידון לא יכתנון מתׄל ובולס שִׁקְמִים. The verb יכׄתנון appears in this edition as יכתנון and is mistakenly translated as "trim [the vines]." As shown by Tirosh-Becker, the phrase ואין מגמזין is part of a citation from the *Mekhilta de-R. Simeon bar Yoḥai*. See further below.

[60] This term in also attested in Muslim and Christian sources from Palestine and Egypt. Tamīmī uses it in a passage cited by Ibn al-Bayṭār in *Traité des simples d'Ibn al-Baïtār de Malaga* (ed. Mohamed al-Arbi al-Khattabī; n. p.: Dar al-Gharb al-Islami, 1990) 105 line 3 s.v. *jummayz*. (For translations and discussions of this passage, see *Relation de l'Égypt* [cited below], 85-86; *Traité des simples, par Ibn el-Beïthar* [trans. L. Leclerc; Paris: Impr. nationale, 1877-83] 364; and Amar, גידולי ארץ־ישראל, 299-300.) Like the Karaites (eleventh century), Tamīmī (tenth century) is from Jerusalem. In Egypt, it is known from a Coptic-Arabic dictionary, where the phrase *ḥattānu l-jummayz* appears as the gloss of the Coptic version's rendering of בולס שקמים; see V. Loret, "Les livres III et IV (animaux et végétaux) de la *Scala Magna* de Schams-ar-Riâsah (1re partie)," *Annales du service des Antiquités de l'Egypte* 1 (1899) 56 nos. 83-84. It survives in modern Egyptian Arabic; see Keimer, "Petits fruits," 51, 57, 75. Not surprisingly, this technical usage of *ḥatana* was not known in Iraq, where the sycomore is not found. Sherira Gaon uses the verb *šaqqaqa* to refer to the gashing of sycomore figs, and Hai Gaon uses *šarraḥa*. ᶜAbd al-Laṭīf al-Baġdādī (*Kitāb al-ʾifādah wa-l-ʾiᶜtibār* [Damascus: Dār Qutaybah, n.d.], 22 line 2; *Relation de l'Égypt* [ed. S. de Sacy; Paris: Imprimerie Impériale, 1810] 83 n. 36) uses *wasama*.

[61] Ibn Ezra, פירושי תרי־עשר, 311.

comment does not give the impression of being based on it. Nor does it seem to be based on the Mishnah, *pace* Simon,[62] since it does not use the verb מנקב (*m. Shebi.* 2.5) or מסטף. Instead it uses משׂרטט, a post-mishnaic verb derived via reduplication from older שׂרט. Since שׂרטט is normally used of scoring parchment, its appearance in this context would be quite unexpected were it not for the use of its Arabic cognate in the Arabic version of Dioscorides' description of the sycomore: *laysa yanḍaju dūna ʾan yušraṭa bi-miḫlabin min ḥadīd* "it does not ripen unless it is slit with a claw of iron."[63] Ibn Ezra may have gotten the idea for this interpretation from Ibn Balʿam, but his formulation shows that Dioscorides influenced his thinking as well.

The Septuagint's Translation of שקמים: The Meaning and Etymology of συκάμινος

Most scholars have recognized that συκάμινα had the meaning "sycomore figs" as well as "mulberries," and that συκάμινος had the meaning "sycomore" as well as "mulberry tree."[64] This is quite clear in the Septuagint from the collocation with κνίζω, which, as we have seen, was a technical term connected specifically with sycomore horticulture. Nevertheless, there are a few dissenting voices. Thus, E. W. G. Masterman writes: "שִׁקְמָה . . . in LXX wrongly tr[d] by συκάμινος . . .

[62] Ibn Ezra, פירושי תרי־עשר, 311.

[63] *La 'materia médica' de Dioscórides* (ed. C. E. Dubler and E. Terés; Tetuán/Barcelona: Emporium, 1953-) 2. 120. This translation reached Spain long before Ibn Ezra's time. Ibn Ezra could have seen the description of the sycomore there or in a derivative pharmacological work, such as Ibn Wāfid's *Al-ʾadwiyah al-mufradah* (Spain, 11th century C.E.). In the transliteration of Ibn Wāfid's work into Hebrew letters (MS Escorial G-II-9, f. 123a bot.), we find וליס ינצ̇ג̇ דון אן ישרט במכלב מן חדיד. The same sentence, in Arabic script, appears in Avicenna's *Qanūn*, book 2, part 2, s.v. *jummayz* (Abū ʿAlī al-Ḥusayn Ibn Sīnā, *Al-Qānūn fi al-ṭibb* [Baghdad: Al-Muthanna Library, n.d.] 285 lines 29-30), but Z. Langermann informs me that it seems to have reached Spain too late for Ibn Ezra to have studied it. This sentence derives ultimately from Theophrastus' description cited above. It is striking that Galil (השקמה, 342) uses the verb שׂרט in translating the latter into modern Hebrew: פירות השקמה אינם מבשילים אלא אם כן שורטו על ידי צפורן ברזל תחילה.

[64] Liddell and Scott, *Lexicon*, 1670 s.v.

'the mulberry.'"[65] R. K. Harrison agrees: "LXX incorrectly Gk. *sykáminos* 'mulberry.'"[66] And L. Zalcman refers to the "error of LXX in rendering *šiqmîm* as συκάμινα (συκάμινος = sycamine, another name for mulberry!)."[67]

This view flies in the face of the evidence. Already in the nineteenth century, C. E. Stowe cited Dioscorides' statement in *De materia medica* that some people use the term συκάμινον for the συκόμορον.[68] He could have added that Theophrastus and Strabo use the term συκάμινος to refer to the sycomore in discussing the flora of Egypt.[69] However, I believe that more needs to be said about this usage.

We appear to be dealing with a dialectal difference between Egyptian Greek and other dialects.[70] That would seem to be the implication of Theophrastus' statement that the συκάμινος peculiar to Egypt "to a certain extent resembles the tree which bears that name in our country," viz., the mulberry.[71] Athenaeus clarifies another aspect of the dialectal difference: "Mulberries (συκάμινα)—Although all other peoples without exception call them by this name, the Alexandrians call them μόρα."[72] These two statements would seem to complement each other: the Alexandrians called mulberries μόρα, because they used the usual term, συκάμινα, for a different fruit. Taken together, they suggest that it is no accident that the Alexandrian translators of the Hebrew Bible (unlike Aquila, Symmachus and Theodotion) never use the term συκόμορος. We may summarize this hypothesis in tabular form:

65 E. W. G. Masterman, "Sycomore, Tree," in *International Standard Bible Encyclopaedia* (ed. J. Orr; Chicago: Howard-Severance, 1915) 5. 2877.

66 R. K. Harrison, "Sycamore; Sycamore Tree," in *International Standard Bible Encyclopedia* (ed. G. W. Bromiley; fully revised ed.; Grand Rapids, Mich.: W. B. Eerdmans, 1979) 4. 674.

67 L. Zalcman, "Piercing the Darkness at *Bôqēr* (Amos VII 14)," *VT* 30 (1980) 254-55 n. 12.

68 C. E. Stowe, "Sycamore," in *Dr. William Smith's Dictionary of the Bible* (ed. H. B. Hackett; New York: Hurd and Houghton, 1870) 4. 3130. Liddell and Scott (*Lexicon*, 1670 s.v.) cite Dioscorides for the same purpose.

69 Theophrastus, *Enquiry*, 1. 290-91 (4.2.1); Strabo, *The Geography of Strabo* (trans. H. L. Jones; Cambridge, Mass.: Harvard University Press, 1949) 8. 148-49 (17.2.4).

70 For the Egyptian branch of the *koine* used by the Septuagint, see E. Tov, "The Septuagint," in *Mikra* (ed. M. J. Mulder; Assen: Van Gorcum, 1988) 180-81.

71 Theophrastus, *Enquiry*, 1. 290-91 (4.2.1).

72 Athenaeus, *Deipnosophists*, 1. 222-23 (2.51).

	Egypt	Greece
mulberries	μόρα	συκάμινα
sycomore figs	συκάμινα	συκόμορα

It appears that Egypt is not the only place where συκάμινος referred to the sycomore. Pseudo-Scylax and Strabo mention a town near Mt. Carmel called Συκαμίνων πόλις, which was already in ruins in Strabo's time.[73] Clearly, this was a city named after its συκάμινος-trees (cf. עיר התמרים in Deut 34:3, Judg 1:16, etc.), but were those trees mulberry-trees or sycomores? H. B. Tristram gives the answer: "Sycaminopolis, near the modern Caiffa, derived its name from the Sycomore fig trees, which still flourish on the ancient site."[74] Additional evidence comes from the toponym שקמנה (*m. Dem.* 1.1),[75] generally identified with Συκαμίνων πόλις, even though there is no independent evidence for this. Thus, συκάμινος had the meaning "sycomore" in at least two of the Mediterranean countries where the sycomore is known to have been cultivated in antiquity.

It is generally agreed that συκάμινος is a Semitic loanword,[76] but no

[73] K. Galling, "Die syrisch-palästinische Küste nach der Beschreibung bei Pseudo-Skylax," *ZDPV* 60 (1937) 79-80, 90; Strabo, *Geography* 7. 274-75 (16.2.27). For other references, see M. Stern, *Greek and Latin Authors on Jews and Judaism* (Jerusalem: The Israel Academy of Sciences and Humanities, 1974-84) 2. 292 n. 27. I am indebted to L. Feldman for this reference.

[74] H. B. Tristram, *Natural History of the Bible* (2nd ed.; London: Society for Promoting Christian Knowledge, 1868) 398. In Egypt, a number of toponyms contain the word for "sycomore" (*nh.t*), either alone or in combination; N. Baum, *Arbres et arbustes de l'Egypte ancienne* (Leuven: Departement Oriëntalistiek, 1988) 24-25. Cf. the Palestinian toponym el-Jummeizeh; E. J. Kraeling, "Two Place Names of Hellenistic Palestine," *JNES* 7 (1948) 200.

[75] So in manuscripts; see משנה זרעים עם שינויי נוסחאות מכתבי יד של המשנה (ed. N. Sacks; Jerusalem: Institute for the Complete Israeli Talmud, 1972-75) 1. 168-69. The form in the printed editions, revived in modern Israel, is שקמונה. Cf. also שמעון השקמוני in the *Sifre* (3x) and in *b. B. Bat.* 119a.

[76] For early discussions of the Semitic origin of the Greek word, borrowed with the plural ending, see H. Lewy, *Die semitischen Fremdwörter im Griechischen* (Berlin: R. Gaertner, 1895) 23. Is the synonym συκόμορος (literally "fig-mulberry") a product of folk etymology, similar to examples like *sparrow-grass* < *asparagus*, *cow cumber* < *cucumber* and *woodchuck* < *otchek*, discussed by R. Anttila (*An Introduction to Historical and Comparative Linguistics* [New York: Macmillan, 1972] 92)?

attempt has been made to be more specific. I propose the following hypothesis to explain the origin of the Greek term. The Greeks first encountered the sycomore while sailing along the coast of Palestine. This tree must have been one of the salient features of the land viewed from a ship, since then, as now, it grew right on the sandy beaches.[77] In the ports, during the Persian period, they learned the Aramaic name, שוקמין, a form known to us from Christian Palestinian Aramaic.[78] With a Greek case ending added, this became συκάμινος. The Greek sailors took this name with them to Egypt, which had sycomores, and to other places which did not. In the latter, they applied it to the mulberry tree, which, as Theophrastus noted, is similar in its leaves and other respects. Thus, the original meaning of the Greek term, "sycomore," was preserved only in places that had sycomores.

Aquila's Interpretation of בולס

Aquila renders בולס with the participle of Greek ἐρευνάω "seek, search, examine."[79] This is the same Greek verb that he uses for חקר in Eccl 12:9, where the targum has בלש, "search."[80] It is also the same Greek verb that LXX uses for חִפֵּשׂ, "search" in Gen 31:35, 44:12, 1 Kgs 20:6, and 2 Kgs 10:23, where Onqelos and Jonathan have בלש. As S. Bochart points out, with a reference to Rashi, this interpretation seems to presuppose comparison of בלס with targumic Aramaic בלש.[81] The latter is used transitively in the targum to 1 Kgs 20:6.

Aquila's interpretation is, first and foremost, a conjecture based on phonetic similarity. Such conjectures are common in the exegesis of all ages when dealing with rare words. F. E. Greenspahn notes that "the assumption that similar consonants interchange" is implicit in the

[77] According to A. Danin, "The Origins of Israel's Sycomores," *Israel Land and Nature* 16 (1990/91), 59, the sycomore is so common on Israeli beaches that the average Israeli associates the two.

[78] See F. Schulthess, *Lexicon Syropalaestinum* (Berlin: G. Reimer, 1903) 214 s.v. שוקמא. The form שקומין at Isa 9:10 is due either either to analogy with the singular absolute or to scribal transposition of two similar letters.

[79] Origen, *Hexapla*, 2. 978.

[80] Origen, *Hexapla*, 2. 404; *The Aramaic Version of Qohelet* (ed. É. Levine; New York: Sepher-Hermon, 1978) 116 col. 3 line 4.

[81] Bochart, *Hierozoicon*, 1. 384 (ed. Rosenmüller, 1. 406).

renderings of the ancient versions.[82] The ancient exegetes frequently ignore contrasts among Hebrew sibilants:

> Thus טוּשׂ is rendered by the Aramaic טוס in both the Targum and Peshiṭta, which also appears to translate עמשׂ as if it were עמס. עסק renders עשׂק in several Aramaic versions. שׂתם is apparently understood as equivalent to סתם by all the ancient versions, and שׂשׂה is clearly equated with שׂסה by Aramaic and Greek translations. Similarly, סרף is treated by the Targum (מיקידא) and Vulgate (*conburet*) as if it were שרף. Aquila treats שׂ like ס when rendering פשׂח as "make lame" (ἐχωλανε), while the Septuagint's translation of שׂקד (ἐγρηγορηθη) is used often for שׁקד. Finally, Aquila's αποθετος for שׂפן may indicate that שׂ and צ were also related.[83]

To this list, one might add the interpretation of חִבֵּשׂ in Job 28:11 as חִפֵּשׂ (sibilants and labials) that Aquila shares with Theodotion.[84]

Most of the examples listed above involve *śin*.[85] The assumption that *śin* interchanges with *šin* (same grapheme) or with *samekh* (same phoneme) is more natural and more common among the ancient exegetes than the assumption that *šin* interchanges with *samekh*.[86] The latter assumption appears in *derashot* of the אל תקרי ("read not X but Y") type,[87] but appears to be rare outside of midrash. The only non-midrashic attestations of this assumption that I know of are in Aquila's interpretation of בלס as equivalent to בלשׁ and his interpretation of פשׁח as equivalent to פסח.[88] Assuming that Aquila is, in fact, unusual in this regard, it is tempting to view this as a reflection of his background.

[82] Greenspahn, *Hapax Legomena*, 52.

[83] Greenspahn, *Hapax Legomena*, 53.

[84] J. Reider, *Prolegomena to a Greek-Hebrew and Hebrew-Greek Index to Aquila* (Philadelphia: n.p., 1916) 92; reprinted in S. Jellicoe, *Studies in the Septuagint: Origins, Recensions and Interpretations* (New York: Ktav, 1974) 327. By coincidence, the Greek rendering is ἐξερευνάω "seek out," almost the same verb as before.

[85] More precisely, *šin* with the realization [s]; see the article cited in the following footnote.

[86] See R. C. Steiner, "*Ketiv-Ḳere* or Polyphony: The שׁ-שׂ Distinction According to the Masoretes, the Rabbis, Jerome, Qirqisānī, and Hai Gaon," in *Studies in Hebrew and Jewish Languages Presented to Shelomo Morag* (ed. M. Bar-Asher; Jerusalem: Bialik, 1996) *151-79.

[87] See the sources cited in Steiner, "*Ketiv-Ḳere*," *156 n. 12.

[88] Compare Aquila's rendering of ויפשחני (Lam 3:11) with his rendering of פסחים (Isa 33:23) and with LXX's rendering of ויפסח (2 Sam 4:4).

Aquila was a speaker of Latin and Greek from a non-Jewish family and a non-Semitic environment (Pontus in Anatolia). Presumably, he did not learn Hebrew and Aramaic until he was an adult. As such, he had only one voiceless sibilant in his native phonemic inventory, and it must have been difficult for him to distinguish *šin* from *samekh*. Indeed, even native Palestinian Jews whose primary language was Greek appear to have had difficulty distinguishing among the Semitic voiceless sibilants, not to mention the laryngeals. This can be seen in a bilingual (Greek-Aramaic) ossuary inscription from first-century C.E. Jericho.[89] The relative importance of the two languages in this inscription is clear: the Greek version appears twice, at the top and on the lid; the Aramaic version appears only once, at the bottom. Hence, when we find the name שלמציון written שלמשיון (and the word אמה "his mother" written המה) in the Aramaic version it is reasonable to suspect Greek influence.

A similar explanation has been suggested by E. Y. Kutscher for the form סלם = שלום, which appears once as a greeting in the Bar Kokhba letters:

> סלם. This form is most surprising. It is true that interchanges of *śin* with *samekh* are found in the Qumran scrolls, as well as the Bar Kokhba letters . . . and also in Mishnaic Hebrew; however, there is no example of an interchange of *šin* with *samekh*. . . .
>
> Perhaps one may suggest the following hypothesis: As mentioned, letters in Greek were found among the letters. One may assume that speakers of Greek did not know how to pronounce the sound *šin*, which is missing in their language, just as they did not know how to pronounce the largyngeals and pharyngeals, which do not exist in their language. . . .[90]

Another factor that may have influenced Aquila's interpretations of בלס and פשה is the fluid orthography current in his time. Among the examples of *šin* written for *samekh* at Qumran, we find פשח for פסח "Passover" (Exod 12:48).[91] I do not mean to suggest that Aquila's text

[89] L. Y. Rahmani, *A Catalogue of Jewish Ossuaries in the Collections of the State of Israel* (Jerusalem: Israel Antiquities Authority, 1994) 244 no. 801.

[90] E. Y. Kutscher, לשונן של האיגרות העבריות והארמיות של בר כוסבה ובני דורו, *Leš* 25 (1960-61) 120-21.

[91] E. Qimron, *The Hebrew of the Dead Sea Scrolls* (HSS 29; Atlanta, Ga.: Scholars Press: 1986) 28-29.

of Lamentations read ויפסחני instead of ויפשחני at 3:11. My point is that spellings like פשח for פסח could have created the impression in Aquila's mind that the two voiceless non-emphatic sibilants of Hebrew ([š] and [s]) were free variants.

The above explanations presuppose that Aquila's mastery of Semitic phonology was somewhat deficient. However, the opposite assumption is also possible. One could assume that Aquila's conjecture was based on a sophisticated awareness that Aramaic *šin* sometimes corresponds to Hebrew *samekh* (and vice versa). Every Jew in Roman Palestine would have known that the term for "synagogue" was (בית) כנישתה (from כנש "gather") in Aramaic but (בית) כנסת (from כנס "gather") in Hebrew.[92] Field notes other examples of Aquila's use of Aramaic to shed light on Biblical Hebrew.[93]

It is difficult to know what Aquila meant by "searching (for) sycomores." Some of the later exegetes who followed in Aquila's footsteps are more explicit. David Qimḥi believed that בולס = בולש is used here in the sense of לוקט "picker." In other words, Amos searched the sycomore trees for figs ready to be picked. More recently, Ashbel has written: "whoever wishes to find a sycomore fig without larvae needs to search (לבלוש ולחפש) among hundreds of figs."[94] Rashi's view of the search is discussed below.

Rashi's Interpretation of בולס and Sycomore Silviculture

As noted above, Aquila's interpretation of בולס reappears in the eleventh century in Rashi's commentary:

> ובולס שקמים—מחפש בשקמים לראות איזה עתו לקוץ כדי להוסיף ענפים ואיזה ראוי לקורות שכן דרך שקוצצין בתולת השקמה. ובולס כמו ובולש אלא שעמוס מגמגם בלשונו שכך אמרו למה נקרא שמו עמוס שהיה עמוס בלשונו וישראל קוראים אותו פסילוס כדאיתא בפסיקתא.[95]

[92] Cf. also המסה-אמשי and, in the opposite direction, השמר-אסתמר.

[93] Origen, *Hexapla*, 1:xxiv.

[94] D. Ashbel, הערות לנבואות עמוס, *Bet Mikra* 25 (1965-66) 106.

[95] Rashi, פרשנדתא והוא פירוש רש״י על נ״ך (ed. I. Maarsen; Amsterdam: M. Hertzberger, 1930-) 1.44-45.

> בולס שקמים — searching among sycomores to see which one's time has come to be cut down in order to add branches and which one is suitable for beams, for that is the practice: the virgin sycomore is cut down. And בולס is like בולש but (the former is written here because) Amos had a speech impediment, for that is what they said: Why was he called Amos? Because he was burdened (עָמוּס) in his speech/tongue, and Israel would call him פסילוס, as it says in the Pesiqta.[96]

It is difficult to believe that Aquila's rather idiosyncratic interpretation occurred to Rashi independently. It seems more likely that Rashi learned of it from Byzantine Jews. D. S. Blondheim and N. R. M. de Lange have shown that Aquila's interpretations were preserved by Greek-speaking Jews until the Middle Ages.[97] Rashi's source may have been his pupil and amanuensis, R. Shemaiah. The latter knew Greek and was familiar with Byzantine coins and the customs of Byzantine Jewry; he may have come from southern Italy.[98]

In resurrecting Aquila's interpretation, Rashi connected it to a tradition, recorded in rabbinic sources and in the introduction to Jerome's commentary on Amos, that Amos had some sort of problem in speaking.[99] For Rashi, the problem was a lisp that caused Amos to pro-

[96] Cf. *Pesiq. Rb. Kah.* 16. This version combines two traditions that are still separate in *Lev. Rab.* 10: the פסילוס tradition and an etymology of the name Amos offered by R. Phinehas. The relationship between the two would seem to be clarified by Jerome's prologue to Amos; see n. 99 below.

[97] D. S. Blondheim, "Échos du judéo-hellénisme," *REJ* 78 (1924) 1-14; N. R. M. de Lange, "Some New Fragments of Aquila on Malachi and Job?" *VT* 30 (1980) 291-94; "The Jews of Byzantium and the Greek Bible," in *Rashi 1040-1990: Hommage à Ephraïm E. Urbach* (ed. G. Sed-Rajna; Paris: Cerf, 1993) 207-8; "La tradition des 'révisions juives' au moyen âge: les fragments hébraïques de la Geniza du Caire," in *"Selon les Septante": Hommage à Marguerite Harl* (ed. G. Dorival and O. Munnich; Paris: Cerf, 1995) 134, 139-40.

[98] See A. Grossman, ר׳ שמעיה השושני ופירושו לשיר השירים in ספר היובל לרב מרדכי ברויאר (ed. M. Bar-Asher; Jerusalem: Academon, 1992) 1. 37. M. Banitt (*Rashi: Interpreter of the Biblical Letter* [Tel-Aviv: Tel Aviv University, 1985] 79-130) claims that a tradition based on the Greek versions underlies most of Rashi's definitions of biblical terms. I am indebted to I. M. Ta-Shma for the latter reference.

[99] See M. Rahmer, "Die hebräischen Traditionen in den Werken des Hieronymus," *MGWJ* 42 (1898) 1-2. (I am indebted to S. Z. Leiman for this reference.) Jerome mentions the speech problem in his prologue to Amos; in the same prologue, a few lines earlier, and in the prologue to Joel, Jerome gives various interpretations of the name, but none of them has anything to do with Amos' being unskilled in speaking. It appears, therefore, that although Jerome knew the פסילוס tradition, he did not know R. Phinehas' etymology of the name עמוס; see n. 96 above.

nounce בולש as בולס. The terms used by the rabbis and Jerome, פסילוס and *imperītus sermone*, may also indicate that they had a lisp in mind.

The term פסילוס, derived from Greek ψελλός "inarticulate," appears in a corrupted form (פסילים) in *y. Ned.* 1.2, 37a and *y. Naz.* 1.1, 51a.[100] There it refers to someone who, unable to pronounce [r] correctly, takes a vow to be a נזיק instead of a נזיר.[101] In all likelihood, the term has a similar denotation in midrashic sources that use it to gloss כבד פה וכבד לשון (Exod 4:10).[102] These sources are no doubt making the same point as other sources, midrashic and non-midrashic, that take כבד פה וכבד לשון as referring to specific consonant groups that Moses was unable to pronounce correctly.[103] The same tradition is reflected in Saadia Gaon's translation of ערל שפתים (Exod 6:12) as אלתג׳ אלפם.[104] Arabic *ʾalṯaġu* means "lisping, substituting one sound for another (e.g., *ṭ* for *q*; *y* or *k* for *l*; *ġ* or *ʾ* for *r*; *ṯ* for *s*)."[105] It is, thus, the exact equivalent of פסילוס, and indeed in one translation from Greek, the phrase *li-man bihī luṯġatun ʾaw ruttatun* "to one who has a *luṯġah* or a

[100] I am indebted to S. Abramson ז״ל for these references.

[101] According to Aristotle (*Problems*, 11.30), ψελλός is used of a speaker who elides sounds, while τραυλός is used of one who pronounces a specific sound incorrectly, e.g, a lisper. Semantically, then, פסילוס would appear to be equivalent to τραυλός rather than ψελλός. J. Duffy writes (e-mail communication, Feb. 24, 1999): "For *traulos* and *psellos* in actual practice there may well have been a certain lack of strictness in their use (e.g. vis-à-vis a definition by Aristotle). As a possible piece of evidence for such an assertion one could cite the Souda (or Suda) Lexicon, a 10th cent. compilation based on earlier sources. For *psellos* it gives three synonyms, namely, a. *asemos* = 'indistinct'; b. *anarthros lalon* = 'speaking inarticulately'; and c. *traulos*. In other words, it regards *psellos* and *traulos* as interchangeable." In any event, neither the Greek evidence nor the Semitic evidence supports the translation "stutterer" for פסילוס in M. Sokoloff, *A Dictionary of Jewish Palestinian Aramaic of the Byzantine Period* (Ramat-Gan, Israel: Bar Ilan University Press, 1990) 440. Aristotle (*Problems*, 11.30) gives a third word with this meaning: ἰσχνόφωνος, referring to a speaker who is unable to quickly add one syllable to another. That word is used of Moses in LXX to Exod 4:10 and 6:30.

[102] See מדרש דברים רבה (ed. S. Lieberman; 2nd ed.; Jerusalem: Shalem, 1992) 134-35.

[103] I owe this insight to S. Z. Leiman, who points to the three sources cited in תורה שלמה (ed. M. Kasher; Jerusalem: n. p., 1927-) 3. 173 note to §42. So too J. H. Tigay, "'Heavy of Mouth' and 'Heavy of Tongue': On Moses' Speech Difficulty," *BASOR* 231 (1978) n. 37. I learned of this discussion from A. Koller while reading the proofs of this monograph.

[104] *Oeuvres complètes de R. Saadia ben Iosef al-Fayyoûmî* [ed. J. Derenbourg; Paris: E. Leroux, 1893-99] 88 l. 19.

[105] M. Ullmann, *Wörterbuch der klassischen arabischen Sprache* [Wiesbaden: O. Harrassowitz, 1970-] 2.189-91.

ruttah" renders τοῖς ψελλοῖς καὶ τοῖς τραυλοῖς.[106] The application of the term *ʾalṯaġu* to Moses predates Saadia. It is found already in a paraphrase of Exod 4:10 by Abū Rāʾiṭa (bishop of Takrit in the early ninth century) and in the *Bayān* of Al-Jāḥiẓ.[107]

Jerome's expression *imperītus sermone* "unskilled in speaking" has been understood in various ways. J. A. Soggin writes that "the great commentator was perhaps judging by the canons of Western rhetoric in his time."[108] However, this suggestion overlooks the rabbinic parallels and the related term that Jerome uses in his commentary to Titus 3:9.[109] There he writes that the Jews "are accustomed to ridicule our lack of skill (*imperītia*), especially in the aspirates and certain letters that should be pronounced with a guttural roughness." This passage shows that the term *imperītia* refers to "lack of skill" in pronouncing the characteristic sounds of Hebrew.

It is difficult to say whether Rashi was the first to connect Amos the בולס with Amos the פסילוס. Certainly neither Jerome nor the rabbis mention בולס in discussing Amos' speech problem. On the other hand, the use of the Greek term פסילוס suggests the possibility that the tradition of Amos' speech defect arose among Greek-speaking Jews as an explanation of Aquila's translation. The label פסילוס, attested as a nickname already in the second century B.C.E. (Josephus, *Vita* 1 §3), may well have been applied to Greek-speaking Jews and/or Christians who were unable to pronounce Semitic sounds not found in Greek; cf. Jerome's complaint about being ridiculed for his pronunciation of Hebrew. One might even speculate that Rashi received Aquila's interpretation from his Byzantine source already tied to the פסילוס tradition. That would explain why Rashi passed up the opportunity to mention the speech defect at Amos 5:11, when interpreting בושסכם as בוססכם.[110]

Those modern scholars who have embraced Rashi's synthesis have usually toned down its midrashic appearance by transforming Amos' lisp into a dialectal feature. M. Rahmer writes: "Näher liegt der Hin-

[106] Ullmann, *Wörterbuch*, 2.190a ll. 40-43. Thus, אַלְתַּג = *man bihī luṯġatun* = ψελλός = פסילוס.

[107] Ullmann, *Wörterbuch*, 2.191a ll. 20-27.

[108] Soggin, *Amos*, 12.

[109] *S. Eusebii Hieronymi Stridonensis Presbyteri commentariorum in Epistolam ad Titum* (*PL* 26; ed. J. P. Migne; Paris: Garnier, 1884) 630.

[110] Other alleged interchanges among the sibilants in Amos are cited by Zalcman, "Piercing," 254 n. 11.

weis darauf, dass Amos sich einiger, wohl dem platteren Volksdialekte angehöriger Wortformen bedient, in denen merkwürdiger Weise gerade der S- und Zischlaut afficirt ist. so בולס für בולש (7, 14), בושס für בוסס (5, 11). . . ."[111] W. R. Harper and S. N. Rosenbaum compare the dialectal peculiarity of the Ephraimites recorded in Judg 12:6.[112]

We come now to Rashi's view of Amos' search. According to Rashi, one of Amos' occupations was searching among sycomore trees for those whose time had come to be cut down. At first glance, this interpretation seems bizarre, and it has found few adherents. Even in the Middle Ages, only Rashi's student, Joseph Qara, accepted it. On closer inspection, however, the view has much to recommend it.

Rashi's view of Amos as a silviculturist is based on rabbinic literature, in which the sycomore appears mainly as a valuable and renewable source of construction beams for roofing, etc. According to Rashi's reading of the sources, the tree did not produce anything else that could conceivably have provided Amos with a source of income.[113] Thus, he had little choice but to view Amos' work with the sycomore in terms of the beams.

In his comment, Rashi refers to the practice, in the tannaitic period, of chopping down the young tree when it was strong enough to regenerate, leaving a stump 24 cm. high (the "sycomore anvil").[114] Every

[111] Rahmer, "Die hebräischen Traditionen," 2.

[112] W. R. Harper, *A Critical and Exegetical Commentary on Amos and Hosea* (ICC 13/1; New York: Scribner, 1905) 8; Rosenbaum, *Amos*, 48, 86-90. Rosenbaum (*Amos*, 89 n. 22) cites Z. Ben-Ḥayyim (עברית וארמית נוסח שומרון [Jerusalem: Bialik/The Academy of the Hebrew Language, 1957-77] 4. 113) in support of the thesis that Ephraimite Hebrew lacked the sound [š]. In fact, Ben-Ḥayyim (עברית, 5. 24; *A Grammar of Samaritan Hebrew* [Winona Lake, Indiana: Eisenbrauns, 2000] 36-37) shows the opposite: due to the merger of **ś* with **š* in Samaritan Hebrew, the sound [š] is far more common there than in Masoretic Hebrew! Moreover, Ben-Ḥayyim (עברית, 5. 24; *Grammar*, 36-37) explicitly denies any connection between this feature of Samaritan Hebrew and the inability of the Ephraimites to pronounce שבלת. On the relation of Samaritan Hebrew to the northern dialect(s) of ancient Israel, see now R. C. Steiner, "*Albounout* 'Frankincense' and *Alsounalph* 'Oxtongue': Phoenician-Punic Botanical Terms from an Egyptian Papyrus and a Byzantine Codex," *Or* 70 (2001) 101 n. 37. For the shibboleth incident, see now J. Blau, בעיות בתורת ההגה והצורות של עברית המקרא: פתרונות קודמים וחדשים in דברי האקדמיה הלאומית הישראלית למדעים 9 (2001) 3-10 and the literature cited there.

[113] Rashi read *m. B. Bat.* 2.13 (*b. B. Bat.* 27b) as implying that the שקמה does not belong to the class of food trees (see Rashi's commentary to *b. B. Meṣ* 109a, *b. Pes.* 53a, and *b. Suk.* 43a; I am indebted to A. Koller and D. Regev for the last two references). See further chapter 5 n. 40 below.

[114] For a detailed analysis of this practice, see M. Kislev, — השקמים אשר בתבליטי לכיש

seven years thereafter, the long straight limbs that grew from its stump were harvested for use as rafters.[115] Sycomore beams (קורות שקמה) from the period have been found at Masada.[116] The sycomore was an ideal source of such beams, thanks, in part, to its "extraordinary regenerative powers."[117] As for the quality of the wood, "its light weight and porous structure made it especially suitable for ceilings."[118] Sycomore beams had an important place in Israel's economy, because they were available locally. This made them less expensive than cedar beams, which had to be imported.

It has long been suspected that at least some of the above was true already in the time of Amos.[119] Until recently, the only basis for this suspicion was Isa 9:9: שקמים גדעו וארזים נחליף "sycomores have been chopped; we shall replace[120] them with cedars." This popular boast of the eighth century reflects the same economic reality as *t. B. Meṣ.* 8.32 and *b. B. Meṣ.* 117b, which deal with the permissibility of replacing a collapsed ceiling of sycomore beams with a ceiling of cedar beams.[121] All of these imply that cedar beams were even more expensive than sycomore beams.

Not long ago, additional evidence for sycomore silviculture in the eighth century was pointed out. According to Kislev, twenty-two "sycomore anvils" with straight, developed beams growing out of them are depicted in the Assyrian reliefs of the siege of Lachish in 701.[122] Z.

זיהוי בראי דברי חז״ל in ירושלים וארץ־ישראל: ספר אריה קינדלר (ed. J. Schwartz, Z. Amar and I. Ziffer; Tel-Aviv: Eretz Israel Museum, 2000), 23-30.

[115] See *m. B. Meṣ.* 9.9, as interpreted by Rashi.

[116] N. Liphschitz and G. Biger, השקמה בישראל בעת העתיקה לפי ממצאים בוטניים מחפירות, *Hassadeh* 72 (1992) 772.

[117] N. Hareuveni, *Tree and Shrub*, 87. These powers are due to the abundance of starch stored in the trunk, branches and roots of the sycomore; see Kislev, השקמים, 26.

[118] M. Zohary, *Plants of the Bible* (Cambridge: Cambridge University Press, 1982) 68.

[119] See, for example, Hepper, *Encyclopedia*, 112: "Although the sycomore is a kind of fig . . . , in Egypt and Palestine during biblical times, it was more important for its timber than for its fruits"; Danin, "Origins," 62: "The practice of planting sycomores, mainly as a source of timber, is several thousand years old." See also n. 135 below.

[120] The other meaning of החליף, also intended here, is "regenerate." It is found in the context of cutting down (גדע, כרת) trees, cf. Job 14:7 כי יש לעץ תקוה אם *יכרת* ועוד *יחליף* "there is hope for a tree; if it is cut, it may yet regenerate" and *m. Abod. Zar.* 3. 7 *גידעו* ופיסלו לשם עבודה זרה *והחליף* נוטל מה *שהחליף* "if he chopped it and trimmed it for idolatrous worship and it regenerated, he may take away what it regenerated."

[121] See J. Feliks, עולם הצומח המקראי (Ramat-Gan: Massada, 1968) 54.

[122] Kislev, השקמים, 23-30.

Amar argues that most of these are olive trees, but even he does not deny that some are sycomores.[123]

Also from the eighth century is ABL 467, an Assyrian letter dealing with timber for construction. Lines 17-23 deal with the transportation of GIŠ *mu-us-ki* GIŠ.ÙR.MEŠ "beams of(?) *mušku*-wood."[124] Thompson identifies *mušku* with Hebrew שקמה, arguing that metathesis is not uncommon when there is an *m* in the word.[125] He could also have noted that the *u*-vowel in the first syllable matches that of Christian Palestinian Aramaic שוקמא, Greek συκάμινος, and perhaps Arabic *sawqam* as well.[126] It is even possible that Akkadian *mušku*, if it was realized [mužgu] or [muzgu] in the Assyrian dialect,[127] is the source of the unattested Aramaic word that yielded Mishnaic Hebrew גמזיות "sycomore figs" and Arabic *jummayz* "sycomore."[128] In addition, the word GIŠ.ÙR = *gušūru* "beam" is equivalent in meaning to Hebrew קורה; indeed, Aramaic כשורא < *gušūru* is used in some targums to translate קורה.[129] Hence the phrase GIŠ *mu-us-ki* GIŠ.ÙR.MEŠ is very

123 Z. Amar, הגידולים החקלאיים על־פי תבליט לכיש, *Bet Mikra* 159 (1998-99) 352-54. I am indebted to M. Kislev for this reference.

124 So CAD s.v. *mušku*; AHw is similar.

125 R. C. Thompson, *A Dictionary of Assyrian Botany* (London: The British Academy, 1949) 321-22. I am indebted to J. Huehnergard for this reference. For metathesis in Akkadian, see *GAG* (including Ergänzungen) §36, where one of the examples is the name of a fruit-tree. For a list of Akkadian-Hebrew cognates in which the order of the corresponding consonants is different, see H. Tawil, "Late Hebrew-Aramaic ספר, Neo-Babylonian sirpu/sirapu: A Lexicographical Note IV," *Bet Mikra* 154-55 (1997-98) 340-41. At least some of these must be the product of metathesis in Akkadian. See also chapter 2 nn. 39 and 44 below.

126 See chapter 3 n. 29 below.

127 It is well-known from loanwords and transcriptions that, in Assyria, Akkadian *š* was realized [s] and *k* was realized [g] in many environments, e.g., *Šarru-kīn* > סרגון and *šaknu* > סגן; see S. A. Kaufman, *The Akkadian Influences on Aramaic* (AS 19; Chicago/London: University of Chicago Press, 1974) 139-41; A. R. Millard, "Assyrian Royal Names in Biblical Hebrew," *JSS* 21 (1976) 4; F. M. Fales, *Aramaic Epigraphs on Clay Tablets of the Neo-Assyrian Period* (Rome: Università degli studi "La Sapienza," 1986) 59-66. This would suggest that *mušku* was pronounced [musgu] or, with voicing assimilation, [muzgu]. Alternatively, since the single Assyrian attestation of the word is written *mu-us-ki*, the pronunciation could be [mušgu] or, with voicing assimilation, [mužgu]. The major difficulty with this theory is the absence of voicing in מסכן < *muškēnu* and נשך < Nusku, but it is not insurmountable.

128 See chapter 2 n. 12 below. If this conjecture is correct, the word for "sycomore" went from the Arabian Peninsula to Israel and (from there?) to Mesopotamia, whence it bounced back to Israel and Arabia, undergoing metathesis twice in the process.

129 See J. Levy, *Chaldäisches Wörterbuch über die Targumim* (Leipzig: Baumgärtner, 1867-68) 393 s.v. כשורא.

close to Mishnaic Hebrew קורות שקמה (*m. Shebi.* 4.5, etc.).[130] Thus, if Thompson's identification is correct, ABL 467 is further evidence for the use of sycomore beams in construction in the ancient Near East.[131] They must have been imported by the Assyrians, since the tree does not grow in Iraq.[132]

We have no reason to believe that sycomore silviculture in Israel began only in Amos' time. Already in David's time, there were sycomores in Israel, in the Shephelah.[133] The importance that David attached to his sycomore groves is seen by his appointment of one Baal-hanan as overseer על הזיתים והשקמים אשר בשפלה, "over the olive

[130] There is, of course, a syntactic difference: the Hebrew phrase exhibits the genitive construction, while, as Thompson (*Dictionary*, 321) saw, the Akkadian phrase exhibits apposition. We must therefore translate: "(as for) the *mušku*-wood, the beams (of/which . . .)." For a translation of the Akkadian phrase based on a different view of the syntax, see G. B. Lanfranchi and S. Parpola, *The Correspondence of Sargon II, Part II: Letters from the Northern and Northeastern Provinces* (State Archives of Assyria 5; Helsinki: Helsinki University, 1990) 209. I am indebted to H. Tawil for this reference.

[131] The only other occurrence of *mušku* recorded by the dictionaries is in a lexical list, which equates *ur-zi-nu* with *mu-uš-ku*, both preceded by the plant determinative Ú; see CAD and AHw s.v. Based on this text, Thompson (*Dictionary*, 321) believes that *urzin(n)u* is "probably the same as . . . (b) (šam)(iṣ)*Mus(š)ku, Ficus sycomorus* L." It appears that, for Thompson, the identification of *urzin(n)u* as *Ficus sycomorus* rests largely on the phonetic similarity between *mušku* and שקמה. It is, therefore, surprising that the dictionaries follow Thompson with regard to *urzin(n)u* (AHw s.v. *urzīnu(m)*: "eine Sykomore?"; CAD s.v. *šimeššalû*: "*urzinnu* . . . sycamore[?]"), while offering no identification for *mušku*. In any event, if the *urzin(n)u*-tree is the sycomore, then the latter was known very early in Mesopotamia, since the term is attested already in Old Babylonian texts.

[132] C. C. Townsend and E. Guest, *Flora of Iraq* (Baghdad: Ministry of Agriculture, 1980) 4/1. 87. I am indebted to J. Huehnergard for this reference. Cf. also Strabo's description of Babylonia (*Geography* 7. 200-201 [16.1.5]): "On account of the scarcity of timber their buildings are finished with beams and pillars of palm wood . . . for, with the exception of the palm tree, most of the country is bare of trees and bears shrubs only."

[133] See 1 Kgs 10:27 ויתן המלך את הכסף בירושלם כאבנים ואת הארזים נתן כשקמים אשר בשפלה לרב, repeated in 2 Chr 1:15 and 9:27. We are not told precisely where these groves were, but the description of the Shephelah in Josh 15:33-44 includes the town of Hagedera. The latter was presumably the home-town of Baal-hanan the Gederite, the man in charge of David's sycomores (1 Chr 27:28). One could argue that just as the cattle in the Sharon were supervised by a Sharonite (1 Chr 27:29), so too the sycomores in the Shephelah were supervised by a local person. If this argument is correct, David's sycomores must have been in the vicinity of Gedera. For a different view, see Galil, השקמה, 314.

trees and the sycomores in the Shephelah" (1 Chr 27:28).[134] David undoubtedly had ambitious construction plans for his new capital, and he needed a cheap, abundant source of wood to realize them. Satisfying this need was no doubt the responsibility of Baal-hanan.[135] He may be compared to Asaph, the "keeper of the king's park" in Nehemiah's time, whose job was to supply "timber for roofing the gatehouses of the temple fortress and the city walls and for the house" of any official who had the authority to demand it (Neh 2:8). In Egypt, too, the sycomore was a source of roof timbers.[136] In that country, artefacts made of sycomore wood have been preserved from the third millennium B.C.E. From the Fifth Dynasty we have dummy vases and a column base and from the Sixth Dynasty we have a coffin; sycomore roots and figs have survived from the Predynastic period.[137] It is clear, then, that Rashi's ideas about the economic role of the sycomore in the biblical period were quite accurate.

[134] For the authenticity and antiquity of the list of David's stewards, see J. M. Myers, *I Chronicles* (AB 12; 1st ed.; Garden City, New York: Doubleday, 1965) 185; T. N. D. Mettinger, *Solomonic State Officials* (ConBOT 5; Lund: CWK Gleerup, 1971) 87; M. Heltzer, המשק המלכותי של דוד המלך לעומת המשק המלכותי של אוגרית, *ErIsr* 20 (1989) 175 and 179 n. 13 and the literature cited there; S. Japhet, *I & II Chronicles* (OTL; London: SCM, 1993), 477-79; and I. Jaruzelska, *Amos and the Officialdom in the Kingdom of Israel* (Poznań: Wydawnictwo Naukowe Uniwersytetu im. Adama Mickiewicza, 1998) 183-84. Japhet notes that many of the names on the list fit very well in the period of David. We may add that the Baalistic name Baal-hanan does too; see M. Noth, *Die israelitischen Personennamen im Rahmen der gemeinsemitischen Namengebung* (Stuttgart: W. Kohlhammer, 1928) 119-21. We may also add that the preposing of the title המלך in the phrase למלך דויד at the end of the list (1 Chr 27:31) may also point to a pre-exilic source; see E. Y. Kutscher, הלשון והרקע הלשוני של מגילת ישעיהו השלמה ממגילות ים המלח (Jerusalem: Magnes, 1959) 340; A. Hurvitz, בין לשון ללשון: לתולדות לשון המקרא בימי בית שני (Jerusalem: Bialik, 1972) 45. Of the other eight instances of המלך דויד in Chronicles, at least five are found in Kings, whereas none of the nine occurrences of דויד המלך in Chronicles are to be found there. For the origin and transmission of such lists, see also N. Na'aman, "Sources and Composition in the History of David," in *The Origins of the Ancient Israelite States* (ed. V. Fritz and P. R. Davies; JSOTSup 228; Sheffield: Sheffield Academic Press, 1996) 170-86. For the existence of crown estate in David's time, see C. Schäfer-Lichtenberger, "Sociological and Biblical Views of the Early State," in *The Origins of the Ancient Israelite States* (ed. V. Fritz and P. R. Davies; JSOTSup 228; Sheffield: Sheffield Academic Press, 1996) 104 n. 71.

[135] Cf. O. Borowski, *Agriculture in Iron Age Israel* (Winona Lake, Ind.: Eisenbrauns, 1987) 128: "It seems that the royal sycamore groves in the time of David were maintained for timber rather than food." I am indebted to J. Huehnergard for this reference.

[136] Gale, et al., "Wood," 340.

[137] Gale, et al., "Wood," 340-41.

CHAPTER 2

בולס: Etymology and Meaning

Bochart's Etymology of בולס

The publication of Samuel Bochart's *Hierozoicon* in 1663 brought the problem of בולס considerably closer to a satisfactory solution. Bochart pointed out that both Arabic and Ethiopian have a word *balas* meaning "fig (fruit or tree)." He noted that in the Ethiopian (Geez) translation of the Bible, this word usually renders תאנה, but in Ps 77:52 (and, one might add, Amos 7:14), it renders שקמה.[1] Bochart suggested that בולס is the participle of a denominative verb, formed from the word for "fig (incl. sycomore fig)." Such a verb, he said, would be comparable to the denominative כורם "vintner" < כרם "vineyard" and to Greek denominatives like συκάζω "to gather or pluck ripe figs" < συκῆ "fig."[2]

Bochart's theory can be supported by other parallels. First there is the participle בוקר, used alongside בולס in Amos 7:14. Bochart himself says earlier that בוקר is "from the word בקר *bakar*, i.e., ox."[3] Indeed,

[1] Bochart, *Hierozoicon*, I. 384 (ed. Rosenmüller, I. 406); cf. also H. Ludolf, *Ad suam Historiam æthiopicam antehac editam Commentarius* (Frankfurt am Main: J. D. Zunner, 1691) 204; A. Dillmann, *Lexicon Linguae Aethiopicae* (Leipzig: T. O. Weigel, 1865) 487; and W. Leslau, *Comparative Dictionary of Geʿez* (Wiesbaden: O. Harrassowitz, 1987) 97.

[2] Bochart, *Hierozoicon*, I. 384-85 (ed. Rosenmüller, I. 406-7).

[3] Bochart, *Hierozoicon*, I. 383 (ed. Rosenmüller, I. 405). See chapter 4 below.

according to Bochart, all three participles used to refer to Amos' occupations are denominatives derived from nouns on the pattern *CaCaC*: בולס is derived from בָּלָס*, בוקר from בָּקָר, and נוקד from נָקָד*.[4]

Another important parallel is the denominative verb *ballasa* "to pick figs" in postclassical Yemeni Arabic; its participle, *miballis* (classical vocalization: *muballis*), refers to "one who picks figs from a tree and sells them in the market."[5] The Egyptian Arabic term *gemamzi*(*a*) "sycomore fig grower(s)," derived from *gimmēz* "sycomore," is also relevant, even though it is not a participle.[6] This word is recorded by Brown and Walsingham, who write: "all the work in connection with the crop [of sycomore figs], including the beating of the tree [with a wooden club, so as to make a ring around the trunk] is done by the 'gemamzia' who buy the year's fruit in advance."[7] "The gashing of the figs is also done by the *gemamzia*."[8]

A related Mishnaic Hebrew denominative participle, מגמזין, has long been known from one witness to a rabbinic text listing agricultural activities forbidden during the sabbatical year.[9] Some scholars have dismissed the form as a scribal error for the form attested in most witnesses, מגזמין,[10] but Tirosh-Becker has discovered a new attestation of מגמזין whose date and place of origin command considerable respect.[11]

[4] For the last denominative, see Bochart, *Hierozoicon*, I. 442-43.

[5] M. Piamenta, *Dictionary of Post-classical Yemeni Arabic* (Leiden: E. J. Brill, 1990-1) 38; I. al-Selwi, *Jemenitische Wörter in den Werken von al-Hamdānī und Našwān und ihre Parallelen in den semitischen Sprachen* (Berlin: D. Reimer, 1987) 44.

[6] It is a broken plural with a *nisba*-ending. It is certainly not accurate to gloss it with an infinitive, as does V. Täckholm, *Faraos blomster* (Stockholm: Generalstabens Litografiska Anstalt, 1969) 51.

[7] Brown and Walsingham, "Sycamore," 11. The beating of the tree, described more fully on p. 10 and explained in the caption to figure 1 (see also Galil, השקמה, 341, last four lines), is also reported in a sixteenth-century source; see Prosper Alpin, *Plantes d'Egypte* (trans. R. de Fenoyl; Cairo: Institut Français d'Archéologie Orientale du Caire, 1980) 21.

[8] Brown and Walsingham, "Sycamore," 10.

[9] MS Vienna at *t. Shebi.* 1.11.

[10] Goldmann, *La figue*, 44; J. Levy, *Neuhebräisches und chaldäisches Wörterbuch über die Talmudim und Midraschim* (Leipzig: F. A. Brockhaus, 1876-89) I. 319.

[11] It appears in a citation from *Mekhilta de-R. Simeon bar Yoḥai* preserved in Yeshuʿah b. Yehudah's commentary to Exod 23:11: ואין מפסלין ואין מצדדין ואין מגמזין. On the last of these phrases, ואין מגמזין, the exegete comments: ירידון לא יכתנון מתל ובולס שקמים "they mean: one may not circumcise [sycomore figs during the sabbatical year],

Thus, it is now more likely that מגמזין is an authentic denominative from Mishnaic Hebrew גמזיות, a noun related to Arabic *jummayz*.[12]

But what is the meaning of גמזיות? Does this term refer to sycomore branches (E. Hareuveni, following the Babylonian Geonim) or sycomore figs (F. Goldmann and E. S. Rosenthal, following the Palestinian Talmud) or both (A. Geiger and S. Lieberman)?[13] Rosenthal's claim that the Geonic interpretation relates to a lexically distinct variant reading current in Babylonia (גואזיות instead of גמזיות) undermines the rationale for the compromise position, but that position still retains a certain attraction, since the sycomore, like other trees of the genus *Ficus*, has specialized leafless branchlets ("panicles") that develop into sycomore figs.[14] Indeed, these figs, or "syconia," are not true fruits at all:

like ובולס שקמים." This comment, written in 11th-century Palestine, is preserved in two manuscripts, one of which may itself be from 11th-century Palestine; see Tirosh-Becker, "Rabbinic Quotation," 382-84 and chapter 1 n. 59 above. Other attestations of the root גמז in the *piel* stem are found in liturgical poetry. In the 11th-century poem אמרר בבכי, one of the סליחות-prayers for the Seventeenth of Tammuz, we find the verbal noun גִּמּוּז rhyming with מזמוז and תמוז; A. Rosenfeld, *The Authorised Selichot for the Whole Year* (London: I. Labworth, 1962) 368. However, it is used there metaphorically, in the phrase לשבר בחורי גמוז (based on Lam 1:15), and so its meaning there is doubly uncertain. Much earlier but badly preserved (in a Geniza fragment) is the attestation of גימוז (again rhyming with תמוז) in the poem איל דכמן בהניסן, one of the שבעתא לטל-prayers composed by Joseph b. Nisan of Shaveh-kiriathaim towards the end of the Byzantine period; see *Ma'agarim: Second Century* B.C.E. *— First Half of the Eleventh Century* C.E. (CD-ROM; Jerusalem: The Academy of the Hebrew Language, The Hebrew Language Historical Dictionary Project, 2001) s.v. גמז. If correctly restored, the poem refers to dew as a therapy (זיר וחבוש, cf. Isa 1:6) for גימוז.

[12] The relationship between these two nouns has been recognized since the Middle Ages; see S. Fraenkel, *Die aramäischen Fremdwörter im Arabischen* (Leiden: E. J. Brill, 1886) 140. Presumably, both are borrowed from Aramaic, even though the Aramaic etymon is not firmly attested. For the possibility that Aramaic *gmz* derives indirectly (via Akkadian) from *šqm*, see chapter 1 above.

[13] E. Hareuveni, גמזיות, *Leš* 11 (1940-41) 39-41; Goldmann, *La figue*, 44; E. S. Rosenthal, בירורי מלים וחילופי נוסח in מחקרי תלמוד : קובץ מחקרים . . . מוקדש לזכרו של פרופ׳ אליעזר שמשון רוזנטל (ed. M. Bar-Asher and D. Rosenthal; Jerusalem: Magnes, 1993) 29-36 (I am indebted to S. Friedman for this reference); A. Geiger, "Bibliographische Anzeigen," *ZDMG* 12 (1858) 363; S. Lieberman, תוספתא כפשוטה: באור ארוך לתוספתא (New York: Jewish Theological Seminary, 1955-88) 1. 360-1.

[14] M. A. Murray, "Fruits, Vegetables, Pulses and Condiments," in *Ancient Egyptian Materials and Technology* (ed. P. T. Nicholson and I. Shaw; Cambridge: University

> The syconium, an inflorescence unique to the genus *Ficus*, is a fleshy branch transformed into a hollow receptacle which bears numerous minute flowers on its inner surface. . . . The true fruits are small drupelets ("seeds") each developing in a female flower inside the syconium.[15]

Thus, the meaning of מגמזין is uncertain. The aforementioned Karaite commentary interpreted it as referring to the gashing of sycomore figs. Modern scholars, to the extent that they have viewed this form as authentic, have generally taken it as referring to removing twigs/branches.[16] As noted above, this interpretation is based on the assumption that גמזיות could refer to twigs/branches, an assumption that has now been called into question by Rosenthal.

Bochart's Etymology and a Postbiblical Survival of בלס "Sycomore Fig"

The theory that בולס is a denominative presupposes that Hebrew had a noun בלס or the like referring to the sycomore fig. G. Hoffmann states the presupposition explicitly: "בָּלָס* war . . . wohl Name der Sykomorenfrucht."[17] But where is the evidence for that noun in Hebrew? The asterisk, indicating that the form is unattested, is an acknowledgment that Bochart's theory is based on an unproven conjecture.

It is safe to say that one of the reasons for the skepticism of Weiss ("the etymology of בולס is problematic") and the other scholars cited

---

Press, 2000) 622. It was E. Hareuveni (גמזיות, 39-41) who first called attention to a possible connection between this peculiarity and the conflicting interpretations of the term גמזיות; however, he himself concluded that the term is properly used only of branchlets (sycomore and carob).

[15] D. Zohary, "Fig," in *Evolution of Crop Plants* (ed. J. Smartt and N. W. Simmonds; 2nd ed.; New York: Wiley, 1995) 367, speaking of the ordinary (Carian) fig. Cf. I. J. Condit, *The Fig* (Waltham, Mass.: Chronica Botanica, 1947) 23.

[16] Geiger, "Anzeigen," 363; Lieberman, תוספתא כפשוטה, 2. 493; Moreshet, לקסיקון, 123. S. Friedman (personal communication) notes that the continuation of Lieberman's discussion (loc. cit.) seems to indicate that this interpretation does not reflect his final conclusion.

[17] G. Hoffmann, "Versuche zu Amos," *ZAW* 3 (1883) 119.

above[18] is the near-universal belief that no trace of such a noun survives in any period of Hebrew. I, too, was under that impression when I began my investigation; I assumed that the original noun had fallen into disuse at some point after the time of Amos. When I consulted the dictionaries of Levy, Jastrow, Ben-Yehudah, and the Academy of the Hebrew Language (on microfiche and CD-ROM), I found nothing remotely relevant under the root בלס.[19]

As my investigation progressed, I was astonished to discover that two of the earliest medieval commentaries to *Seder Zeraim* of the Mishnah tell a very different story.[20] The Yemenite expansion of the eleventh-century commentary of Nathan Av ha-Yeshivah[21] to *m. Maas.* 2.8 reads: "בלסים: inferior figs known as *khanas* (sycomore)."[22]

[18] See chapter 1 above.

[19] Levy, *Wörterbuch*; M. Jastrow, *A Dictionary of the Targumim, the Talmud Babli and Yerushalmi, and the Midrashic Literature* (London: Luzac, 1903); Eliezer Ben-Yehuda, מלון הלשון העברית הישנה והחדשה (New York/London: T. Yoseloff, 1960); *Materials for the Dictionary, Series I: 200 B.C.E. — 300 C.E.* (Jerusalem: The Academy of the Hebrew Language, The Historical Dictionary of the Hebrew Language, 1988); *Ma'agarim: Second Century* B.C.E. *— First Half of the Eleventh Century* C.E. (CD-ROM).

[20] The readings of most, but not all, of the textual witnesses cited below are recorded in משנה זרעים עם שינויי נוסחאות, 2. 192, 219.

[21] Two manuscripts of the expansion are extant: MS British Library Or. 11117 (G. *Margoliouth, Catalogue of the Hebrew and Samaritan Manuscripts in the British Museum* [London: The British Museum, Dept. of Oriental Printed Books and Manuscripts, 1899-1935] 4. 161) and MS JTS R 1492. For the history of the latter manuscript, see M. Z. Fox (H. Fox), תשלום המלקט מפירוש רב נתן אב הישיבה למשנה, in לראש יוסף: מחקרים בחכמת ישראל (ed. J. Tobi; Jerusalem: Afikim, 1995) 371-86. Genizah fragments of the original commentary have been identified by N. Danzig and H. Fox; unfortunately our passage is not preserved in them (oral communication from H. Fox). One of those fragments, originally called to my attention by E. Hurvitz, has been published by Fox; see M. Z. Fox (H. Fox), המשנה בתימן; כתוב־יד מפירוש רב נתן אב הישיבה, *Asufot* 8 (1994) 161-67. It shows that the original commentary consisted of short glosses to individual words, much like אלפאט̇ אלמשנה "words of the Mishnah" and מילי קושיאתא "difficult words" treated in N. Allony, מחקרי לשון וספרות (Jerusalem: Ben Zvi Institute, 1986) 1. 137-98, esp. 172-73. Indeed, the remnant of the name of the work preserved on the fragment—מיל[י]—confirms that it belongs to the genre of lexical commentaries on the Mishnah.

[22] MS British Library Or. 11117, f. 21: הבלסים תין דני יערף באלכ̇נס. JTS R 1492 f. 65a reads הבלסים תין דני יערף באלבלס. Both *khanas* "sycomore" and *balas* "fig" are peculiar to the dialect of Yemen. Assuming that the words יערף באלכ̇נס were added by the editor, it is clear that he must have been a Yemenite (Qafiḥ's view), not an Egyptian

And the commentary of Isaac b. Melchizedek of Siponto (c. 1090-1160) to *m. Ter.* 11.4 reads: "בליסים: a species of fig."[23] According to these commentaries, the noun בלס *does* survive in the Mishnah, in several places. One commentary knows that it refers to a specific variety of fig but is unable to identify it; the other identifies the variety as *khanas*, the Yemeni Arabic term for *Ficus sycomorus*. These commentaries are not the only witnesses attesting to the existence of this form. Although most manuscripts read לבסין or the like in *m. Maas.* 2.7-8, the citation of this passage in *y. Maas.* 2.7, 50a has the form בלוסים (probably from בליסים), at least according to the Leiden manuscript and the first edition (Venice, 1523).[24] Similarly, while most manuscripts have כליסים "pods" or the like in *m. Ter.* 11.4, a Babylonian Genizah fragment (TS E,128) reads בליסים, and R. Solomon Sirillo reports that (his version of) the Arukh, the famous dictionary of R. Nathan of Rome, had the reading בליסין.[25]

(Assaf's view); see Fox, המשנה בתימן, 165. On the other hand, it is likely that the words הבלסים תין דני come from the original 11th-century commentary, which is comprised of short lexical glosses; see the preceding footnote.

[23] בליסים מין תאינין הן. This is the reading of both of the extant manuscripts: MS British Library Or. 6712, f. 111a and MS Oxford (Neubauer) 392; the latter is printed in Isaac b. Melchizedek of Siponto, פירוש הריבמ״ץ לרבנו יצחק ב״ר מלכי צדק מסימפונט למשנה זרעים (ed. N. Sacks; Jerusalem: Institute for the Complete Israeli Talmud, 1975) 272. In theory, the form בליסים could be a segolate plural exhibiting vowel harmony in the stem (**CiCīC-* instead of **CiCaC-*) comparable to Mishnaic Hebrew נְזִיקִין (sg. נֶזֶק-), Biblical Hebrew פְּסִילִים (sg. פֶּסֶל-), נְטִעִים (sg. נֶטַע-), Arabic *kisirāt* "fragments" (sg. *kisrah*), *sidirāt* "lotus trees" (sg. *sidrah*), etc.; see W. Wright, *A Grammar of the Arabic Language* (3rd ed.; Cambridge: University Press, 1896-98) 1. 192. However, as J. Huehnergard notes (e-mail communication, Jan. 30, 2002), this would presuppose that the noun had the otherwise unattested form **bils-*. Accordingly, it seems preferable to accept the suggestion of D. Talshir (e-mail communication, Aug. 19, 2001) that we are dealing with a different noun pattern, similar to that of Arabic *jummayz*. Support for this suggestion comes from the (Babylonian) vocalization בַּלִֵיסִים in TS E,128; see I. Yeivin, מסורת הלשון העברית המשתקפת בניקוד הבבלי (Jerusalem: The Academy of the Hebrew Language, 1985) 2. 1066.

[24] For the readings of these and other witnesses, see תלמוד ירושלמי יוצא לאור על פי כתב יד סקליגר 3 (Or 4720) שבספריית האוניברסיטה של ליידן עם השלמות ותיקונים (Jerusalem: The Academy of the Hebrew Language, 2001) 272 lines 2-3 and *Synopse zum Talmud Yerushalmi* (ed. P. Schäfer and H.-J. Becker; Tübingen: J. C. B. Mohr, 1992) 1/6-11. 180-1, bottom.

[25] See Solomon b. Joseph Sirillo, מסכת תרומות מן תלמוד ירושלמי . . . לר׳ שלמה בכ״ר יוסף סיריליאו (ed. P. Shapiro and J. Freimann; Benei Berak: הרשקוביץ, 1958) 159b.

Sycomore figs fit nicely in the aforementioned passages from the Mishnah. *M. Maas.* 2.7-8 deals with limitations on the right of a worker to eat the figs (תאנים) that he was hired to pick. While working with an inferior variety (called בלסים by Nathan Av ha-Yeshivah and לבסין or the like by most other witnesses), he may not eat a superior variety (בנות שבע)[26] and vice versa. There is no difficulty in assuming that the term תאנים in Mishnaic Hebrew—like the terms תאנים in medieval Hebrew,[27] תאנֵא in Syriac,[28] *tīn* in Arabic,[29] *fici/ficus* in Latin,[30] and *figs* in English[31]—could be used in a broad sense to include sycomore figs. And sycomore figs were certainly considered inferior in the time of the Mishnah.[32]

M. Ter. 11.4 (and *m. Uqṣin* 1.6) contains a list of fruits that reads, according to Isaac b. Melchizedek of Siponto and the Babylonian Genizah fragment: תאינין וגרוגרות והבליסים והחרובין. Here again we have the collocation with תאנים "figs," this time augmented by גרוגרות

[26] According to *y. Bik.* 1.3, 63d, this is the preferred variety of figs for the first fruit offering, since it is מן המובחר "among the choicest." According to one opinion in Gen. Rab. 15 (מדרש בראשית רבא [ed. J. Theodor and C. Albeck; Jerusalem: Wahrmann, 1965] 140-1), the irresistible fruit eaten by Adam and Eve (like the leaves that they subsequently wore) were from a fig tree of the ברת שבע variety.

[27] Estori Farḥi (כפתור ופרח [ed. A. M. Luncz; 3rd ed; Jerusalem: A. M. Luncz, 1899] 527 lines 2-3) glosses בנות שקמה with תאנים מדבריות "wild figs" and תאני פרעה "Pharaoh's figs."

[28] Bar Serošwai (c. 900 C.E.) *apud* Bar Bahlul (*Lexicon*, 2005 lines 14-15) defines שְׁקְמא as tasteless תאנֵא, which the Arabs call *jummayz*. Among the definitions given by Ishodad of Merv in his commentary to Amos 7:14 (*Commentaire* [CSCO 303] 4. 90 lines 11-12), we find תאנֵא דברא "wild figs."

[29] *Book of Plants* by ʾAbū Ḥanīfa al-Dīnawar *apud* ʿAbd al-Laṭīf al-Baġdādī, *Kitāb al-ʾifādah wa-l-ʾiʿtibār*, 22 lines 15-16: *wa-min ʾajānisi t-tīni tīnu l-jummayzi wa-huwa tīnun ḥulwun raṭbun....* "One of the kinds of fig is the sycomore fig, which is a sweet juicy fig. . . ." Maimonides (משנה, 1. 132) glosses בנות שקמה with אלג׳מיז והו איצׄא תין ברי "the sycomore, which is also a wild fig."

[30] According to Jerome (*Commentarii in Prophetas Minores*, 324 line 402), the *sycamina* bear *agrestes . . . ficus* "wild figs."

[31] See Condit, *Fig*, 3.

[32] In *t. Ter.* 5.7, גמזיות, "sycomore figs," are mentioned among the fruits whose תרומה "the priests do not care about." Strabo (*Geography*, 8. 148-49 [17.2.4]) reports that the fruit "is not prized for its taste." The Palestinian Talmud (*y. Dem.* 1.1, 21c) implies that the owners of sycomore trees normally abandoned the figs; cf. chapter 1 n. 53 above. See also chapter 5 n. 40 below.

"dried figs." The collocation with חרובין "carob fruit" is reminiscent of the frequent pairing of the sycomore tree with the carob tree in rabbinic literature.[33] Theophrastus too puts the two together.[34]

It should therefore come as no surprise that the Yemenite editor of the commentary of Nathan Av ha-Yeshivah uses the Yemeni Arabic term *khanas* "sycomore"[35] to gloss כליסים as well as בלסים and that Maimonides takes כליסים to be a kind of fig. A. Kohut, too, takes the word as referring to a type of fig, and emends to בלוסין, while Sirillo emends to לובסין based on his text of *m. Maas.* 2.8.[36]

The term כליסים also occurs in *t. Ter.* 5.6-7, where the context is seemingly less favorable to such emendations. According to *t. Ter.* 5.6, only one sixtieth of the harvest need be given as תרומה in the case of certain produce that the priests did not care about. The only examples given are הכליסין והחרובין. In 5.7, a longer list of such produce is given: הקצח והכליסין והחרובין וגמזיות ותרמוסין ושעורין אידומיות. Since גמזיות means "sycomore figs,"[37] emending הכליסין to הבליסין would seem to be precluded—not only in *t. Ter.* 5.6-7 but also in *m. Ter.* 11.4 and *m. Uqṣin* 1.6. However, it is possible that the list in *t. Ter.* 5.7 is a composite, reflecting two different traditions, since the first three items have the definite article, while the last three do not. The term for sycomore figs may have been בליסין in one tradition and גמזיות in the other. It is

[33] See the many examples cited in Goldmann, *La figue*, 42 n. 8 and add *m. B. Bat.* 2.13, 4.8, 9; *b. Pes.* 56b; *b. Men.* 71b.

[34] After his description of two varieties of sycomore (the "Egyptian sycamine" and the "Cyprian fig"), Theophrastus continues (*Enquiry*, 1.294-95 [4.2.4]): "Like this too is the tree which the Ionians call carob; for this too bears most of its fruit on the stem . . . ; some call it the 'Egyptian fig'—erroneously; for it does not occur at all in Egypt, but in Syria and Ionia and also in Cnidos and Rhodes."

[35] JTS R 1492 f. 131a הכליסים אלכ̇נס והו אלתאלוק וקיל תין ברי. *Tālūq* is another Yemeni botanical term, taken here to be a synonym of *khanas* = *Ficus sycomorus*; see also Piamenta, *Dictionary,* 52. However, *tāluq* is identified by P. Forskål (*Flora Ægyptiaco-Arabica* [Hauniæ: Ex officina Mölleri, 1775] 179 and CXXIV) as the Yemeni term for *Ficus vasta*. Forskål's identification is accepted by G. Schweinfurth, "Sammlung arabisch-æthiopischer Pflanzen," *Bulletin de l'Herbier Boissier* 4 (1896) Appendix II, 129-31; A. Al-Hubaishi and K. Müller-Hohenstein, *An Introduction to the Vegetation of Yemen* (Eschborn: Deutsche Gesellschaft für Technische Zusammenarbeit, 1984) 162, 196; and P. Behnstedt, *Die nordjemenitischen Dialekte* (Wiesbaden: L. Reichert, 1985-92) 2. 138.

[36] Nathan b. Yeḥiel, ערוך השלם, 4. 241a s.v. כלס; Sirillo, מסכת תרומות, 159b.

[37] See at nn. 13-14 above.

also possible that Mishnaic Hebrew בלס referred not to the sycomore fig but to an inferior variety of the common (Carian) fig.[38]

Despite all of the uncertainties, it is clear that the noun בלס survives in Mishnaic Hebrew (if not in its original form, then certainly in the metathesized form לבס),[39] referring to the sycomore fig or some other sort of inferior fig. One may surmise that, if it referred specifically to the *sycomore* fig, it was already an archaic literary term by the time that the Mishnah was edited, supplanted in colloquial usage by גמזיות "sycomore figs" and the more transparent term בנות שקמה "fruits of the sycomore."[40] That would help to explain why, in most manuscripts of the Mishnah, there is no trace of בל(י)סים. Instead, as noted above, we find forms like לבסין in *m. Maas.* 2.7-8 and כליסים in *m. Ter.* 11.4, *m. Uqṣin* 1.6, and *t. Ter.* 5.6-7.[41] It is thus understandable that the form בלס does not appear as the name of a type of fig in the four major dictionaries of Mishnaic Hebrew cited above.[42] However, it is less understandable that the metathesized form, לבס, found in 14 of 17 manuscript witnesses to *m. Maas.* 2.8,[43] not to mention the most common edition of the Mishnah (Vilna, 1908), is also unrecorded in these dictionaries or else buried in another entry as a variant reading.[44]

[38] For the varieties of the common fig in the Near East, see Condit, *Fig*, 22-23.

[39] It is not impossible that the language actually had both בלס and לבס. For metathesis involving *lamed*, cf. שמלה-שלמה, בהלה-בלהה, Ezra 4:4 Kt מבלהים Qr מבהלים, etc. For metathesis from the same semantic field, cf. אלגומים-אלמגים and n. 44 below. The antiquity of the form לבס would be confirmed if *HALAT* s.v. בלס, were right in identifying the Egyptian *nbs*-tree with the sycomore, but see chapter 3 n. 60 below.

[40] For בנות "daughters of" used in the sense of "fruit of (a tree)," see Goldmann, *La figue*, 44 on Syriac בנת ארזא, בנת אסא, etc. and Y. Feliks, עצי־פרי למיניהם: צמחי התנ״ך וחז״ל (Jerusalem: R. Mass, 1994) 200 on בנות הדס = בנת אסא, בנות שבע, בנות חריע. It is probably an Aramaism.

[41] Some sources even have קליסין instead of כליסים: Codex Parma B (De Rossi 497) at *m. Uqṣin* 1.6; פירוש הגאונים לספר טהרות (ed. J. N. Epstein; Jerusalem/Tel-Aviv: Magnes/Devir, 1982) 140.

[42] See n. 19 above.

[43] So according to the files of התלמוד הישראלי השלם. I am indebted to Rabbi J. Hutner for granting me access to these files. The summary in משנה זרעים עם שינויי נוסחאות, 2. 219, cites fewer manuscripts.

[44] One must turn to a concordance of the Mishnah to find an entry for לבס, referring to a type of fig; see C. Y. Kasovsky, אוצר לשון המשנה (Tel-Aviv: Massadah, 1967) 2. 1043. It is totally absent from the dictionaries of Ben-Yehudah and the Academy of the Hebrew Language. In Levy, *Wörterbuch*, 236, it appears s.v. בלופסין. In Jastrow, *Dictionary*, 640, it is listed s.v. כלופסין. It should be noted that Jastrow's gloss, "Lesbian fig,"

The survival of this word in rabbinic literature but not in the Bible should not be surprising. It has long been recognized that a number of ancient Hebrew botanical and agricultural terms—nouns and verbs omitted by chance from the Bible—are preserved only or mainly in rabbinic literature.[45] Among the nouns are שׂרף "sap, resin"; זרדים "shoots" (cf. the toponym נחל זרד); שחליים "cress" (cf. Old Aram. שחלין, Aram. תחלין, Akkad. *saḫlû*); כרישים "leeks" (cf. Aram. כרתין, Arab. *kurrāṯ, karrāṯ*, Akkad. *karašu, karšu*); צמל "ripe fig."[46]

Two of these terms belong to the same semantic field as בלס. The word שׂרף "sap, resin"[47] is used in phrases like שׂרף הפגים, שׂרף התאינה, שׂרף השקמה (*m. Orlah* 1.7, *t. Miq.* 6.9) to refer to the milky sap (latex) of the fig and the sycomore.[48] There is no reason to doubt the conventional assumption that this term, derived from a good Hebrew root (שׂרף "burn"),[49] was in use during the biblical period. If the use of latex

---

is unfounded. The form לבס has nothing to do with the Greek island of Lesbos; it is derived from בלס via metathesis; see n. 39 above. Curiously, the word for inferior *grapes*, באשים, also undergoes metathesis in Mishnaic Hebrew, yielding אבשים.

[45] E. Y. Kutscher, מלים ותולדותיהן (Jerusalem: Kiryath Sefer, 1965) 28, 79-81, based on Löw, *Flora*, passim; A. Sáenz-Badillos, *A History of the Hebrew Language* (trans. J. Elwolde; Cambridge: Cambridge University Press, 1996) 200. It is generally accepted today that Mishnaic Hebrew was descended from a colloquial idiom spoken in the biblical period; see the survey of the literature in R. C. Steiner, "A Colloquialism in Jer 5:13 from the Ancestor of Mishnaic Hebrew," *JSS* 37 (1992) 11-26. The importance of Mishnaic Hebrew for biblical lexicography, has been stressed by R. Gordis ("Studies in the Relationship of Biblical and Rabbinic Hebrew," in *Louis Ginzberg Jubilee Volume* [New York: American Academy for Jewish Research, 1945] 173-75); J. C. Greenfield ("Lexicographical Notes I," *HUCA* 29 [1958] 204); B. A. Levine ("Survivals of Ancient Canaanite in the Mishhah" [Ph.D. diss., Brandeis University, 1962] 2; I am indebted to M. S. Smith for this reference); and E. Y. Kutscher ("Mittelhebräisch und jüdisch Aramäisch im neuen Köhler-Baumgartner," in *Hebräische Wortforschung: Festschrift zum 80. Geburtstag von Walter Baumgartner* [VTSup 16; Leiden: Brill, 1967] 158-59), among others.

[46] For the first three examples, see Kutscher, מלים, 79-80. I am indebted to J. Huehnergard for calling the Akkadian cognates to my attention.

[47] Vocalized שֶׂרֶף or שְׂרָף in reliable manuscripts.

[48] For fig latex, see Condit, *Fig*, 25. For sycomore latex, see Keimer, "Petits fruits," 61. Keimer specifically notes that this latex "coule . . . de presque toutes les parties de *Ficus sycomorus*, de l'écorce quand on l'incise, des feuilles, des fruits. . . ." Similarly, the Mishnah speaks of latex of the leaves and latex of the roots (שׂרף העלים, שׂרף העקרים) as well as latex of the figs (שׂרף הפגים) in *m. Orlah* 1.7.

[49] The root may allude here to the sensation caused by a proteolytic enzyme found in

to make cheese (*m. Orlah* 1.7) goes back to the biblical period, as the evidence of Homer (*Iliad* 5.902-3) and other Greek writers suggests,[50] the term probably does too. The antiquity of Mishnaic Hebrew צמל "ripe fig" was argued by J. C. Greenfield, who compared it to Ug. *ṣml*:

> The word *ṣemel* is a virtual hapax in Mish. Heb. and is preserved in the sort of comparison that has all the marks of an earthy folk tradition. This lends strength to the assumption that we are dealing with an ancient term that may very well be "Canaanite" in origin.[51]

The evidence presented in this chapter confirms Bochart's etymology in a rather conclusive manner. The etymology was inspired by a recognition of the relationship between the Hebrew participle בולס and the Arabic/Ethiopian noun *balas*. However, Bochart's picture was incomplete, as shown by the empty cells in the following chart:

	noun	denominative participle
Hebrew	—	בולס
"Arabic," Ethiopian	*balas*	—

---

latex called "ficin." According to Condit (*Fig*, 25), "this enzyme accounts for the dermatitis often experienced by some packers of dried figs and especially by pickers and consumers of fresh figs." Even latex gloves can cause a reaction.

[50] See Löw, *Flora*, 1. 246-7; Condit, *Fig*, 25; Liddell and Scott, *Lexicon*, 1241 s.v. ὀπός. It has been suggested that this use of latex may go back to prehistoric times; see J. M. Renfrew, *Palaeoethnobotany: The Prehistoric Food Plants of the Near East and Europe* (New York: Columbia University Press, 1973) 136. Ficin (see the preceding footnote) is still used "in the cheese industry as a substitute for rennet in the coagulation of milk" according to *The Merck Index: An Encyclopedia of Chemicals, Drugs, and Biologicals* (Rahway, N.J.: Merck and Co., 1989) 4019. I am indebted to J. Crystal for this reference.

[51] J. C. Greenfield, "Ugaritic Lexicographical Notes," *JCS* 21 (1967) 90; cf. also his "Amurrite, Ugaritic and Canaanite," in *Proceedings of the International Conference on Semitic Studies held on Jerusalem, 19-23 July 1965* (Jerusalem: The Israel Academy of Sciences and Humanities, 1969) 99. J. Huehnergard calls my attention to the fact that others are much less certain of the meaning of Ug. *ṣml*; see G. del Olmo Lete and J. Sanmartín, *Diccionario de la lengua ugarítica* (Aula Orientalis-Supplementa 8; Sabadell, Barcelona: Editorial AUSA, 1996-2000) 2. 418 s.v. *ṣml* (III). Note also that, *pace* Greenfield, the vocalization found in reliable manuscripts is צַמָּל.

We have now found the data to close these gaps:

	noun	denominative participle
Hebrew	לבס/בלס	בולס
Yemeni[52] Arabic	*balas*	*miballis*

As a result, there is no longer room for doubts about Bochart's etymology. The term בולס is the participle of a denominative verb, derived from the noun בלס. The implications of this etymology, which are weightier than Bochart realized, are the subject of the next section and the next chapter.

Bochart's Etymology and the Meaning of בולס שקמים

What are the semantic ramifications of Bochart's etymology? Is it compatible with LXX's interpretation ("a scratcher of sycomores")? Bochart himself abandoned that interpretation after examining a group of Greek verbs "formed from various fig names" that "signify whatever pertains to their care."[53] He concluded that בולס refers to *qui ficus colit, & sycaminos* "one who tends fig and sycomore (trees)."[54]

Bochart does not go so far as to claim that his etymology is incompatible with the LXX's interpretation, but Greenspahn does: "It seems unlikely that the denominative of a kind of tree would refer to such an isolated part of its treatment."[55] The translation of the LXX, he says, "could be based on an educated guess relying on the botanical practice with regard to such trees in the translators' environment."[56] Greenspahn's point would be well taken if the verb בלס were, in fact, derived from a word denoting a tree. It would then be natural to expect that a

[52] See chapter 3 below.

[53] Bochart, *Hierozoicon*, 1. 384-85 (ed. Rosenmüller, 1. 407).

[54] Bochart, *Hierozoicon*, 1. 385 (ed. Rosenmüller, 1. 407). The same conclusion was reached in the Middle Ages by Al-Qumisi (פתרון, 38) and Isaiah of Trani (פירוש, 98).

[55] Greenspahn, *Hapax Legomena*, 106.

[56] Greenspahn, *Hapax Legomena*, 106.

בולס would be responsible for all aspects of the work involved with that tree, like a כורם "vintner" and a *gemamzi*. However, we have shown above that בולס is derived not from the name of the sycomore *tree* but from the name of its *fruit*.[57]

Another argument that has been raised against the Septuagint's interpretation is independent of Bochart's etymology. Goldmann writes that those who believe that a בולס שקמים is a gasher of sycomore figs "oublient que ce ne peut être un métier."[58] W. Rudolph develops this argument: "Auf alle Fälle handelt es sich um eine Arbeit, die im Ablauf der jährlichen landwirtschaftlichen Verrichtungen nur wenig Zeit in Anspruch nimmt, so daß es unwahrscheinlich ist, daß sie als Berufsbezeichnung diente (kein deutscher Weingärtner wird als seinen Beruf 'Rebenspritzer' angeben). . . ."[59] That the gashing of sycomore fruit is not even close to being a full-time job is confirmed by the observations of Brown and Walsingham: "The work extends over a period of two to three days for each crop."[60] It should be noted, however, that other common interpretations of בולס שקמים are open to the same objection.[61]

The Septuagint's interpretation has had a number of defenders in modern times. Some of them (Keimer, Galil) may have been unaware of Bochart's etymology,[62] but others (G. Baur, Hitzig, Lagarde, P. Humbert) have seen no inconsistency in accepting Bochart's etymology while rejecting his interpretation.[63] G. Baur writes:

> . . . so ist auch das Verbum בָּלַס ein Denominativum dieser Art. Ihm die allgemeinste Bedeutung, "Maulbeerfeigen bauen" (Gesenius u. d. W.) zu geben, geht nicht an, weil nach dem Verbum noch einmal שקמים steht, der hebräische Ausdruck wäre dann ebenso ein Pleonasmus, wie wenn

[57] Cf. also the conjecture of Baur quoted below.

[58] Goldmann, *La figue*, 45 n. 1.

[59] W. Rudolph, *Joel—Amos—Obadja—Jona* (KAT 13/2; Gütersloh: G. Mohn, 1971) 257.

[60] Brown and Walsingham, "Sycamore," 10.

[61] I am indebted to J. Huehnergard for this observation.

[62] Keimer, "Bemerkung," 441-42; Galil, השקמה, 347.

[63] Hitzig, *Kleinen Propheten*, 142-43; P. de Lagarde, "Ueber die semitischen Namen des Feigenbaums und der Feige," in *Mittheilungen* (Goettingen: Dieterichsche Sortimentsbuchhandlung, 1884-91) 1. 59. The latter refers to "das hebräische Denominativum בלס *ein caprificierender* Amos 7, 14." The term *caprificierender* is used here in an extended sense, as explained by Reynier ("Méthode," 184-89); Löw (*Flora*, 1. 275) criticizes that use of the term.

wir im Deutschen sagen wollten: "Gras grasen"; vielmehr da שקמה den ganzen Baum bezeichnet . . . , בלס nach dem Kamus die Frucht insbesondere, so muss das Verbum eine mit dieser insbesondere vorgenommene Thätigkeit bezeichnen, wie in unserm "den Weinstock beeren", "das Kraut blättern" u. dergl. Am nächsten liegt nun, es durch "Maulbeerfeigen vom Baume sammeln" zu erklären; aber schon der ganz eigenthümliche Ausdruck deutet auf einen minder allgemeinen Begriff, und aller Wahrscheinlichkeit nach haben die LXX . . . das Richtige getroffen. . . .[64]

P. Humbert suggests an ingenious combination of Bochart's etymology with LXX's interpretation. He argues that Arabic *balas* refers only to a *ripe* fig and that its denominative therefore means "to ripen a fig (artificially)." His source is the definition given by *Lisān al-ʿArab*: *ʾal-balasu ṯamaru t-tīni ʾiḏā ʾadraka* "*balas* is the fruit of the fig-tree when it has ripened." He concludes that "בָּלַס nicht bloss, wie bisher angenommen wurde, mit Feigen zu tun haben bedeutet, sondern die Reife der Feigen (oder Maulbeerfeigen) befördern."[65] However, most sources do not support the notion that the meaning of the noun is restricted to *ripe* figs.[66]

In my view, neither the interpretation of the Septuagint nor the interpretation of Bochart is precisely on the mark. Bochart's interpretation rests to a great degree on his comparison of בולס with כורם, but that comparison, although very insightful, is imprecise in two respects. First, בולס is derived from the name of a fruit, while כורם is derived from the name of a tree or, rather, a collection of trees. In other words, we are not dealing with a denominative *שוקם comparable to כורם and *gemamzi*. What we have is בולס comparable to συκάζω and probably *miballis*, both referring to the harvesting of fruit.

Second, בולס is transitive, while כורם is intransitive. In other words, בולס is part of a phrase (בולס שקמים) that has far more in common with

[64] G. Baur, *Der Prophet Amos* (Giessen: J. Ricker, 1847) 411-12.

[65] P. Humbert, "בּוֹלֵס שִׁקְמִים (Amos VII, 14)," *OLZ* 20 (1917) 296-98. A partial parallel exists in Modern South Arabian. Jibbāli has a noun *ḥfɔl* meaning "ripe wild fig" and a number of verbs from the same root; T. M. Johnstone, *Jibbāli Lexicon* (Oxford: Oxford University Press, 1981) 104-5. The intransitive verbs have meanings like "to ripen, to be ripe, to be ripe enough to eat" (all restricted to wild figs), but transitive *aḥfel* means "to collect wild figs."

[66] Indeed, one source defines Yemeni *balas* as *ṣigāru t-tīn* "young figs"; Zayd ibn ʿAli ʿInān, *Al-Lahja al-yamāniyya fī al-nukat wa-l-ʾamṯāl al-ṣanʿāniyya* (n.p.: 1980) 45 no. 460.

Lev 19:10 וכרמך לא תעולל (and Baur's "den Weinstock beeren") than with כורם, assuming that *BDB* is right in taking תעולל as a denominative from עוללות. In both cases, a verb derived from the name of a fruit takes the name of the tree as its object. Here too the parallel refers to the harvesting of fruit.

Thus, the closest parallels to בלס support the interpretation of the Peshiṭta, David Qimḥi, and Abarbanel: לקט "pick, gather."[67] It follows that Amos' term בולס שקמים referred to a person that harvests the fruit of the sycomore. It may have stood in contrast to some term for a person who used the sycomore tree as a beam factory, something like מגדע/גוזר/כורת שקמים, "a cutter of sycomores." This may explain why Amos called himself a בולס שקמים rather than a שוקם*. The latter term would not have been capable of distinguishing the two very different modes of exploiting the sycomore tree in ancient Israel—the horticultural and the silvicultural. These two modes may well have been incompatible to a certain extent, for the treatment that promotes the growth of straight, smooth limbs inhibits the growth of fruit.[68] In later Hebrew, the difference between the two modes is reflected in terms for the tree itself: a sycomore that has been cut to produce beams is called a "sycomore anvil" (סדן השקמה); a sycomore that is allowed to grow naturally (and produce figs) is called a "virgin sycomore" (בתולת השקמה). Kislev has argued that sycomores of both types are depicted in the Assyrian reliefs of the siege of Lachish in 701, not long after Amos' time.[69]

Galil believes that this conclusion can be undermined by asking why Hebrew needed a special verb for picking sycomore figs in addition to the verb לקט "pick, gather":

> There are simple and accepted words in the Hebrew language to describe growing, gathering or picking. Why was this strange word, found nowhere else in the Bible, used here, if it is not the description of an action that is *sui generis*, connected exclusively with the fruit of the sycomore?[70]

[67] So too AV: "a gatherer of sycomore fruit." As noted above, Qimḥi combines this contextual interpretation with an old etymology that equates בלס with Aramaic בלש "search."

[68] Kislev, השקמים, 26, 28.

[69] Kislev, השקמים, 23-30. See chapter 1 above.

[70] Galil, השקמה, 347.

There are two ways of dealing with this argument. One is to point to other verbs for gathering or picking in Hebrew that are restricted to a single kind of fruit, e.g., בצר for grapes and Mishnaic Hebrew מסק for olives. Another is to point out that accepting the Peshiṭta's interpretation does not necessitate total rejection of the LXX's interpretation.

According to the latter approach, these two ancient interpretations are not as far apart as they seem. The gashing of the sycomore figs may well have been viewed as the beginning of the harvest, since the picking of the figs followed only a few days later. Theophrastus writes that "the fruits thus scratched ripen in four days," while Athenaeus says "they become ripe and fragrant in three days."[71] Reynier says that the gashed figs reach maturity "en peu de jours."[72] Galil reports that "in three to four days the figs increase about seven times in weight and volume."[73] According to Brown and Walsingham: "The work extends over a period of two to three days for each crop, and the fruit is ready for gathering four or five days after the holes have been made."[74] Keimer too writes that "a few days later the fruit is picked."[75]

Even stronger evidence could be adduced from the reports of Sickenberger and Figari, if only they were credible. According to the former:

> The operation is only made on fruits which shall be picked up the following day. The day after the operation the fig is quite ripe.[76]

According to the latter, the gashing was done after the figs were cut from the tree.[77]

We conclude that בלס refers to the entire process of harvesting sycomore figs, beginning with the gashing. The distinctiveness of that initial step may have promoted the coining of a special verb for the whole process.

[71] Athenaeus, *The Deipnosophists*, 1. 222-23.

[72] Reynier, "Méthode," 188.

[73] Galil, "Ancient Technique," 186.

[74] Brown and Walsingham, "Sycamore," 10.

[75] Keimer, "Knife," 65.

[76] *Apud* Henslow, "Egyptian Figs," 102; revised version in Henslow, "The Sycomore Fig," 129. Sickenberger's report appears to be based largely on Figari's.

[77] Figari, *Studii*, 2. 177-78.

CHAPTER 3

בולס and שקמה: Linguistic Evidence for the Origin of the Biblical Sycomore

The Controversy Surrounding the Origin of the Biblical Sycomore

The conclusions of the preceding chapter can help to resolve the controversy surrounding the origin of the biblical sycomore. It has generally been assumed that the sycomores in the Shephelah were already there when David ascended the throne.[1] But how did this tropical tree get there, and when? Discussions of these questions in the seventies and eighties were inconclusive.

In 1976, Galil, Stein, and Horovitz asked: "Has the sycomore fig moved to the Middle East spontaneously as a part of the wild Sudanian element and remained in the area after losing its reproductive capacity only because of the active help of man, or has man been entirely responsible for its northward transportation?"[2] After much

[1] Japhet (*I & II Chronicles*, 478) writes: "The acquisition and origin of this vast agricultural property — land, vineyards, groves of olives and sycamore, cattle, sheep, camels and asses — are nowhere documented. The incidental geographical terms, Shephelah, Sharon and 'the valleys', probably indicate that these estates came to David as a result of his wars, when he lay claim to the royal property of conquered Canaanite states and cities, but this conclusion should be adopted only with caution." See also Heltzer, המשק, 177.

[2] J. Galil, M. Stein & A. Horovitz, "On the Origin of the Sycomore Fig (*Ficus sycomorus* L.) in the Middle East," *The Gardens' Bulletin, Singapore* 29 (1976) 192.

deliberation, they were unable to reach a unanimous verdict.[3] In 1985, Galil himself was still unable to make up his mind:

> The disparity between the sycomore's rhythm of activity and the climatic rhythm in Israel, the absence of its pollinators, and its dependence on humans for propagation are seemingly evidence that it was brought by humans and planted here many years ago. But a second possibility exists—that the sycomore, like the other wild tropical plants that grow here, reached Israel on its own. . . . It is difficult to determine which of these two possibilities is correct.[4]

M. Żohary seemed to have a definite opinion on the matter in 1982:

> Some scholars assume that the species was introduced from Africa, perhaps by Natufian man (about 10 000 BC) bringing seeds or cuttings. . . . In my opinion, it was never in fact "introduced" into Israel, but remained as a tertiary relic of the other tropical flora, not unlike other vestiges (*Acacia albida*, *Ziziphus spina-Christi*). . . .[5]

However, in a Hebrew encyclopedia entry published in the very same year, he gave a different picture: "It is assumed that the sycomore was introduced (הוכנסה) into the land of Israel from Egypt, and that it was brought to the latter from East Africa in the third millennium B.C.E."[6] In more recent literature, a consensus appears to be emerging, with writers describing the sycomore as having been "imported," "brought," or "introduced" into Israel.[7]

As for the date, there are at least four theories: (1) the "Canaanite period," i.e., the Bronze Age;[8] (2) the end of the Neolithic Period;[9]

[3] Galil, Stein & Horovitz, "Origin," 202.

[4] J. Galil, הפיקוס: עץ בר ועץ נוי (Jerusalem: משרד החינוך והתרבות, 1985) 77.

[5] M. Zohary, *Plants*, 68.

[6] M. Zohary, שקמים, in אנציקלופדיה מקראית (Jerusalem: Bialik, 1972-88) 8. 258. Volume 8 appeared in 1982.

[7] The first term is used by Liphschitz and Biger (השקמה, 771) in presenting the "accepted" view; the second is used by Kislev (השקמים, 24); the third is used by D. Zohary and M. Hopf, *Domestication of Plants in the Old World* (3d ed.; Oxford: Clarendon Press, 2000) 164-65. Cf. also the article by Danin cited below.

[8] Galil, הפיקוס, 67; Kislev, השקמים, 24.

[9] Galil, הפיקוס, 77 (as an afterthought).

(3) the Natufian (Mesolithic) Period;[10] and (4) the Tertiary Period.[11] Naturally, these two controversies are not unrelated: the earlier the arrival, the less likely it was to have involved human agency.

The three early datings (2-4) find no support in the fossil and archeological records. No sycomore fossils have been found in Israel.[12] The earliest remains of *Ficus sycomorus* discovered in Israel are from the Iron Age.[13] The lack of earlier finds is striking, because fossilized leaves of *Ficus carica* embedded in travertine rock have been discovered at En Gedi,[14] and "charred fig pips have been retrieved from numerous early Neolithic sites in the Near East . . . such as PPNA [pre-pottery Neolithic A] (7900-7500 bc) Netiv Hagdud, Israel . . . , PPNA (7000 bc) Jericho . . . , aceramic Neolithic (7800-6600 bc) Tell Aswad, Syria . . . , and PPNB (7200-6000 bc) ʿAin Ghazal, Jordan. . . ."[15] The absence of *Ficus sycomorus* remains is particularly striking at Jericho, because remains believed to be *Ficus carica* (carbonized pips and, in one instance, flesh of the fruit) were found there in many of the oldest layers (not only PPNA, but also Chalcolithic, Early Bronze, and Middle Bronze)[16] and because Jericho was well known for its sycomores in later times.[17] It is not surprising, then, that Galil, Stein and Horovitz are forced to admit that "evidence for the presence of *F. sycomorus* in the Middle East in ancient times is far from satisfactory, especially since the data supporting the presence of the plant in Natufian Palestine are based on an indirect method of inquiry."[18]

[10] Galil, Stein & Horovitz, "Origin."

[11] M. Zohary, *Plants*, 68 (see at n. 5 above and the background given in M. Zohary, *Plant Life of Palestine* [New York: Ronald Press, 1962] 61, 63-64).

[12] Cf. Baum, *Arbres*, 21, arguing against M. Zohary.

[13] Personal communication from M. Kislev. For sycomore wood excavated in Israel from the Iron Age on, see Liphschitz and Biger, השקמה, 772. The reference there to sycomore wood from PPNA Jericho must rest on some sort of misunderstanding.

[14] Danin, "Origins," 61.

[15] D. Zohary and Hopf, *Domestication*, 163.

[16] M. Hopf, "Plant Remains and Early Farming in Jericho," in *The Domestication and Exploitation of Plants and Animals* (ed. P. J. Ucko and G. W. Dimbleby; Chicago: Aldine, 1969) 356-57.

[17] See chapter 5 below. In Deuteronomy and Judges, however, Jericho is עיר התמרים, not עיר השקמים.

[18] Galil, Stein & Horovitz, "Origin," 202.

The case for migration of the sycomore to Israel without human intervention is even weaker, as A. Danin has demonstrated:

> Israel's sycomores were obviously nurtured by human effort. They develop tasty, juicy "figs," which do not contain fruits. Thus, they cannot spread locally through seeds—nor could they have germinated from fruits coming from a different area.
>
> The sycomore fruits closest to Israel originate in southern Sudan. This is far too distant for the fruit bats or the birds which spread sycomore seeds around. Even if viable seeds were to be transported somehow, they could not germinate and develop under the conditions of Israel's southern coastal plan.
>
> Tropical trees growing in Israel require an abundance of water during the hot seasons.[19] Even the Mediterranean maquis bushes, which develop in the northern coastal plain, and germinate during winter and spring, cannot grow in the drier southern part.
>
> The sycomores of Israel were thus obviously planted by farmers. . . .[20]

The import theory, on the other hand, has much to recommend it. The practice of transplanting trees from abroad is attested early among Israel's neighbors. Long before David's time, Egyptian monarchs like Hatshepsut brought myrrh trees from Punt and planted them in the temple compounds.[21] The Assyrian kings, too, were active in this regard:

> Tiglath-Pileser I in the eleventh century is the first Assyrian king who reveals in his inscriptions a clearly utilitarian interest in establishing

[19] The Hebrew version (A. Danin, השקמה אינה עץ בר, *Teva Vaaretz* 32 [1990] 31) adds: "The sands in the south of the coastal plain never have enough water for . . . germination, and certainly not in the hot season of the year." The point is that sycomore seeds require even more water than cuttings.

[20] Danin, "Origins," 62. As for the theory that the sycomore grew spontaneously in Upper Egypt in the Predynastic period (see n. 64 below), Danin writes: "I think that with the extreme desert areas located between the two areas (S. Egypt and Israel) and the missing habitats for both *F. sycomorus* and birds & bats, the seeds could not cross the "barrier" by natural agents; humans could transport it and plant it in places where no trees can establish themselves" (e-mail communication, Sept. 14, 2002).

[21] F. N. Hepper, "An Ancient Expedition to Transplant Living Trees," *Journal of the Royal Horticultural Society* 92 (1967) 435-38; D.M. Dixon, "The Transplantation of Punt Incense Trees in Egypt," *JEA* 55 (1969) 55-65. I am indebted to L. E. Stager for these references.

> gardens outside his capital at Assur to cultivate foreign trees for timber and fruit. He records that he "took cedar (*erēnu*), box-tree (*taskarinnu*), Kanish oak (*allakanish*) from the lands over which I gained dominion [in the west]—such trees which none among previous kings, my forefathers, had ever planted—and I planted [them] in the orchards of my land. I took rare orchard fruit which is not found in my land [and therewith] filled the orchards of Assyria." . . .
>
> In the ninth century Assurnasirpal II, in the remarkable text on the so-called Banquet Stela, recorded the trees, seedlings, and plants that he had seen on his military campaigns and then imported for planting in the irrigated gardens that he had created in his new capital at Kalah (Nimrud). Here pleasure and utility were blended. The range of species is wide, many at present untranslatable, extending from trees like cedar/pine, cypress, and juniper valued for constructional timber, to fruit and ornamental trees and shrubs. . . . Oppenheim has suggested that it was Sargon II in the later seventh century BC who changed the motivation of the royal patrons of gardening 'from utilitarian to display purposes, from an interest in assembling the largest possible variety of specimens to incorporating a garden into the palace precinct for the personal pleasure of the king. . . .[22]

In the following sections, we shall see additional evidence, of a linguistic and archeological nature, that supports this theory.

The Distribution of *Bls* and *Šqmt* in the Semitic Languages

The full significance of Bochart's etymology has not been seen, even by those who accepted the conjecture that Hebrew had a noun בלס, because it has been assumed that that noun and the Arabic-Ethiopian noun *balas* were independent reflexes of a common Proto-West-Semitic ancestor. However, this assumption is difficult to maintain once we compare the geographical distribution of *bls* with that of *šqmt*.

[22] P. R. S. Moorey, *Ancient Mesopotamian Materials and Industries* (Oxford: Clarendon Press, 1994) 349. See also D. J. Wiseman, "Mesopotamian Gardens," *Anatolian Studies* 33 (1983) 138. It appears that Solomon continued this tradition; for this and other parallels between Solomon and the kings of Assyria, see B. Halpern, "The Construction of the Davidic State: An Exercise in Historiography," in *The Origins of the Ancient Israelite States* (ed. V. Fritz and P. R. Davies; JSOTSup 228; Sheffield: Sheffield Academic Press, 1996) 48-51.

Based on what we know today, we can say that Bochart's characterization of the distribution of *bls* was in some ways too narrow and in others too broad. In Arabia, the word is peculiar to the Yemeni dialect.[23] In Ethiopia, on the other hand, the word is found not only in Geez, but also in Amharic, Tigre, Tigrinya, Gurage, Argobba, and Soddo.[24] In Hebrew, as we have shown, the noun is attested as well as the denominative participle. From a geographic perspective, the distribution of the word is rather limited. It is found only in Ethiopia, Yemen and Israel. It is not recorded for the Modern South Arabian languages spoken in Dhofar (Oman),[25] despite the presence of the sycomore and other Ficus species there.[26]

The word *šqmt* has a similar distribution, although one would not know it from *HALAT*. The only cognates of Hebrew שקמה listed by *HALAT* are Syr. שקמא and Christian Palestinian Aramaic שוקמא > Greek συκάμινος. No mention is made there of the Arabic forms cited in the nineteenth-century dictionaries of W. Gesenius, A. Kohut and T. Audo. Gesenius compares שקמה with colloquial Arabic *sokam*, a rare name of the *Ficus sycomorus*.[27] The form *sokam* comes from Forskål,

[23] Cf. Našwān bin Saʿīd al-Ḥimyarī, *Šams al-ʿUlūm* (ed. K. V. Zetterstéen; Leiden: E. J. Brill, 1951-53) 185 line 21: *al-balasu t-tīnu bi-luġati l-yaman* "*balas* are figs (*tīn*) in the language of Yemen." See also al-Selwi, *Jemenitische Wörter*, 44; C. von Landberg, *Glossaire Daṯinois* (Leiden: E. J. Brill, 1920-42) 1. 204. It is used not only of *Ficus carica* (the common fig) but also of *Ficus palmata* Forsk.; see Forskål, *Flora*, CXXIV; Schweinfurth, "Sammlung," 125, 128; Al-Hubaishi and Müller-Hohenstein, *Introduction*, 196.

[24] See Leslau, *Comparative Dictionary*, 97; idem, *Etymological Dictionary of Gurage (Ethiopic)* (Wiesbaden: O. Harrassowitz, 1979) 3. 142; W. M. Kelecha, *A Glossary of Ethiopian Plant Names* (4th ed.; Addis Ababa: n.p., 1987) 48, 119. In addition to *Ficus carica* (the common fig), it is used of *Ficus palmata* Forsk. (Amharic, Tigre and Tigrinya) and *Ficus capreæfolia* Del. (Tigre); see Kelecha, loc. cit. and Schweinfurth, "Sammlung," 125.

[25] Jibbāli has a number of words for wild figs and wild fig trees; see Johnstone, *Lexicon*, 92 s.v *ġyź*; 104-5 s.v. *ḥfl*; 282 s.v. *ṭyḳ*.

[26] D. Heller and C. C. Heyn *(Conspectus Florae Orientalis: An Annotated Catalogue of the Flora of the Middle East* [Jerusalem: The Israel Academy of Sciences and Humanities, 1994] 28 and map 1) include Dhofar in the distribution of *F. sycomorus*, *F. palmata* and *F. vasta*. I am indebted to J. Huehnergard for this reference. According to A. G. Miller and T. A. Cope (*Flora of the Arabian Peninsula and Socotra* [Edinburgh: Edinburgh University Press, 1996-] 94-102, 506-7), Oman is home to those species and also *F. carica*, *F. johannis*, *F. cordata*, and *F. ingens*.

[27] W. Gesenius, *Thesaurus philologicus criticus linguae Hebraeae et Chaldaeae Veteris Testamenti* (Leipzig: F. C. W. Vogel, 1835) 1477.

who also supplies an unpointed Arabic transcription: سقم.[28] Combining the two transcriptions, we get *soqam*, with two short vowels. Kohut and Audo compare שקמה with classical Arabic *sawqam*, defined as a fig-like tree in the dictionaries.[29]

Even Gesenius, Kohut and Audo omit the crucial information that this Arabic word originated in Yemen. As noted by Rabin, *sawqam* is labeled a "Yemeni expression" (*luġah yamāniyyah*) by Ibn Durayd.[30] And Forskål heard the word *soqam* in the Yemen Highlands or the nearby plain, somewhere between Al-Luḥayya and Taᶜizz, in 1762-63.[31] In 1881, Schweinfurth recorded *súggama* as a vernacular name of *Ficus sycomorus* in Ḥaḍramawt, at al-Hāmī, east of al-Shiḥr on the southern coast of the Arabian Peninsula.[32] The *g* of this form represents the

[28] Forskål, *Flora*, CXXIV.

[29] Nathan b. Yeḥiel, ערוך השלם, 5. 150a s.v. שקמה; T. Audo, *Sīmtā d-leššānā sūryāyā* (Mossoul: Imprimerie des pères dominicains, 1897) 596b. Is *sawqam* a hypercorrection for *soqam*? Such a hypercorrection could have been promoted by the existence of a *fawᶜal* pattern in Arabic. I am indebted to W. P. Heinrichs for this latter suggestion.

[30] C. Rabin, *Ancient West-Arabian* (London: Taylor's Foreign Press, 1951) 27. Rabin's gloss of Ibn Durayd's *sawqam* is "sycamore," which agrees with later Yemeni usage but not with Ibn Durayd's own description. Ibn Durayd (*Kitāb jamharat al-luġah* [ed. R. M. al-Baᶜlabakkī; Beirut: Dār al-ᶜIlm lil-Malāyīn, 1987-88] 2. 851 col. 2 lines 2-3) says that the tree resembles (but is not the same as) the *ḫilāf*. The *ḫilāf* is *Salix aegyptiaca* L., the Egyptian willow; Maimonides ("Moses Maimonides' Glossary of Drug Names," in *Maimonides' Medical Writings* [Haifa: The Maimonides Research Institute, 1995] 311-12) actually gives *sālij*, an Arabic rendering of the Old Spanish reflex of *salix*, as one of its alternate names. M. Kislev suggests that Ibn Durayd may be referring to *Ficus salicifolia* (personal communication). As its name implies, the leaves of this very widespread relative of the sycomore resemble those of the willow; see S. Collenette, *Flowers of Saudi Arabia* (London: Scorpion, 1985) 369 and Miller and Cope, *Flora*, 100-101, 507. Kislev's suggestion appears to be confirmed by other dictionaries, which state that the *sawqam* is "exactly like the *ʾaṯʾab*, which is a tree of the fig-kind . . . having a fruit like the fig . . ."; E. W. Lane, *Arabic-English Lexicon* (London: Williams and Norgate, 1863-77) 1384. Schweinfurth ("Sammlung," 133) reports that *Ficus salicifolia* is called *athâb* (and *thaâb*) in Yemen. Even if Schweinfurth's *athâb* represents *ʾaṯᶜab* (cf. *ṯaᶜb* in Al-Hubaishi and Müller-Hohenstein, *Introduction*, 196 and *ṯᶜb* in Forskål, *Flora*, CXXIV), it seems likely that it is a colloquial form or by-form of *ʾaṯʾab*. For ʾ > ᶜ in Arabic, see S. Fraenkel, *Beiträge zur Erklärung der mehrlautigen Bildungen im Arabischen* (Leiden: E. J. Brill, 1878) 12-13. (I am indebted to J. Blau for this reference.)

[31] Forskål, *Flora*, CXXIV. The abbreviations there are explained on p. CI. The expedition is described on pp. LXXXVI-XC.

[32] G. Schweinfurth, "Sitzungs-Bericht vom 15. October 1889," *Sitzungs-Berichte der Gesellschaft naturforschender Freunde zu Berlin* (1889) 158 and Schweinfurth, *Sammlung*, 143. The latter gives the form as *sugguma*.

"bedouin" realization of *q*; the final *a* represents the feminine ending, which converts mass nouns into count nouns (*nomina unitatis*).

It seems likely that Epigraphic South Arabian (henceforth ESA) is the source of both *balas* (Yemen and Ethiopia) and *soqam* (Yemen),[33] and, in fact, the latter is attested in Qatabanian, although this has not been recognized. The term *s*^{1}*qmtm* occurs twice in an inscription from Wadi Bayḥān (RES 4932): *s*^{2}*hr ġyln bn ʾbs*^{2}*bm mlk qtbn bny ws*^{1}*ḥd[ṯ brdʾ ʿṯtr nwp]n wʾlhw s*^{1}*qmtm bytn byḥn . . . ywm rdʾ ʿṯtr wʾlhw s*^{1}*qmtm s*^{2}*hrġln mḥḍ ḥḍrmt wʾmrm. . . .* RES translates: "Šahr Ġaylân, fils de ʾAbšibâm, roi de Qatabân, a bâti et renouvelé [avec l'aide de ʿAthtar Nawfâ]n et des divinités d'irrigation, le temple Bayḥan . . . lorsque assurèrent ʿAthtar et les divinités d'irrigation à Šahr Ġaylân la défaite de Ḥaḍramût et ʾAmrum."[34] The phrase *ʾlhw s*^{1}*qmtm*, translated "les divinités d'irrigation," is similarly rendered by Ryckmans ("irrigation deities")[35] and Ricks ("irrigation gods").[36] This rendering of *s*^{1}*qmtm* can hardly be correct: the word for "irrigation" in Qatabanian and elsewhere in ESA is *ms*^{1}*qt*, *ms*^{1}*qyt* from the root *s*^{1}*qy*.[37] A different interpretation of *ʾlhw s*^{1}*qmtm*, based on Arabic *saqima* "be sick," is given by Jamme: "les divinités de la maladie?" or, more precisely, "les divinités invoquées en temps de maladie."[38] Jamme does not explain why such gods would not be called "gods of healing" or

[33] So too Yemeni *jafn* "grapevine"; see Selwi, *Jemenitische Wörter*, 63 and J. C. Biella, *Dictionary of Old South Arabic* (HSS 25; Chico, Calif.: Scholars Press, 1982) 74 s.v. *gfn*. For South Arabian loanwords in the Arabic of Yemen, see Rabin, *Ancient West-Arabian*, 26; cf. also 45-47. For Sabean influence on Ethiopian Semitic vocabulary, see D. L. Appleyard, "Ethiopian Semitic and South Arabian: Towards a Re-examination of a Relationship," *Israel Oriental Studies* 16 (1996) 208.

[34] RES VII, 434-35.

[35] *Apud* F. Stark, "Some Pre-Islamic Inscriptions on the Frankincense Route in Southern Arabia," *JRAS* (1939) 497. See also G. Ryckmans, "Inscriptions sub-arabes; cinquième série," *Le Muséon* 52 (1939) 66-67 no. 216.

[36] S. D. Ricks, *Lexicon of Inscriptional Qatabanian* (Rome: Editrice Pontificio Istituto Biblico, 1989) 10 s.v. *ʾL* and 162 s.v. *S*^{1}*QM*.

[37] Ricks, *Lexicon*, 162 s.v. *S*^{1}*QY*. In J. Huehnergard's unpublished notes on RES 4932, which he was kind enough to send me after reading this discussion, the translation "irrigation" for *s*^{1}*qmtm* is labeled "v[ery] unlikely."

[38] A. Jamme, "Le panthéon sud-arabe préislamique d'après les sources épigraphiques," *Le Muséon* 60 (1947) 120 n 558. In Jamme's view ("Le panthéon," 89-90, 124-27), the ESA term for "irrigation god" is *mnḍḥ*. Ricks (*Lexicon*, 110 s.v. *NḌḤ*) takes that term to mean "tutelary deity."

the like. Nor does he explain why they would be responsible for assuring the defeat of neighboring countries.

I suggest that the term *ʾlhw s¹qmtm* should be interpreted in the light of the Hittite term *DINGIR*^MEŠ (*LÚ*^MEŠ) ^GIŠ*ERIN-aš* "cedar-gods."[39] If so, it means "sycomore-gods," i. e., the gods who dwell on/in/under the sycomore(s).[40] Already in the pyramid texts of ancient Egypt (3rd millennium B.C.E.), we find a reference to "yonder tall sycamore in the east of the sky . . . on which the gods sit."[41] In those texts, there is also a reference to gods *under* a sycomore, and from later periods there are numerous representations of individual goddesses (Nut, Nut/Hathor, Isis, Nephthys, Neith) *in* a sycomore.[42] There was a sycomore tree in the courtyard of the Eleventh Dynasty temple of Mentuhotpe at Deir el-Baḥri, judging from the roots discovered there by Winlock.[43] Indeed, one scholar has speculated that every temple garden in ancient Egypt had a sycomore.[44] According to *t. Abod. Zar.* 7.(7) 8, at least one sycomore in postbiblical Palestine

[39] L. Zuntz, "Un testo ittita di scongiuri," *Atti del Reale Istituto Veneto di scienze lettere ed arti* 96/2 (1936-37) 488-526, 530-31; *ANET*, 351-53; B. H. L. van Gessel, *Onomasticon of the Hittite Pantheon* (Leiden: Brill, 1998) 992-93. The signs *LÚ*^MEŠ, used to write this expression in *KUB* XV but not in *KBo* VI, indicate the masculine nature of the deities, according to Zuntz, *Testo*, 531.

[40] The initial sibilants of ESA *s¹qmtm* and Hebrew שקמה correspond regularly; see A. F. L. Beeston, "On the Correspondence of Hebrew *ś* to ESA *s*²," *JSS* 22 (1977) 50.

[41] R. O. Faulkner, *The Ancient Egyptian Pyramid Texts* (Oxford: Clarendon Press, 1969) 159 §916; cf. R. Moftah, *Die heiligen Bäume im Alten Ägypten* (Ph. D. diss., Georg-August-Universität zu Göttingen, 1959); idem, "Die uralte Sykomore und andere Erscheinungen der Hathor," *Zeitschrift für ägyptische Sprache und Altertumskunde* 92 (1966) 40-47; H. Kees, *Der Götterglaube im alten Ägypten* (Berlin: Akademie-Verlag, 1987) 84 and P. Koemoth, *Osiris et les arbres* (Liège: C.I.P.L., 1994) 55-56, 59-60, and passim.

[42] Baum, *Arbres*, 44, 67-86.

[43] A. Lucas, *Ancient Egyptian Materials and Industries* (4th ed.; London: Histories and Mysteries of Man, 1989) 447.

[44] G. Schweinfurth, "Über die Bedeutung der 'Kulturgeschichte,'" *Botanische Jahrbücher* 45 (1910), Beiblatt 103: 34. For the Egyptian temple gardens, see Dixon, "Transplantation," 59; J.-C. Hugonot, *Le jardin dans l'Egypte ancienne* (Frankfurt am Main: P. Lang, 1989) 32, 35, 40, 41, 58, 65, 67, 75; *idem*, "Ägyptische Gärten," in *Der Garten von der Antike bis zum Mittelalter* (ed. M. Carroll-Spillecke; Mainz am Rhein: P. von Zabern, 1992) 33-38; A. Wilkinson, *The Garden in Ancient Egypt* (London: Rubicon, 1998) 119-44. A famous Egyptian sycomore that is venerated to this day is the Virgin's Tree in North Cairo; see S. Sachs, "A Tree Drooping with its Ancient Burden of Faith," in *The New York Times*, Dec. 26, 2001, A4.

was venerated as a sacred tree: שלש אשירות בארץ ישראל חרוב שבכפר קסם ושבכפר פגשה ושקמה שבראני/ו ושבכרמל "There are three idolatrous trees in the Land of Israel: the carob in Kefar Qsm and that in Kefar Pgšh and the sycomore in Rʾny/w and that at Carmel."[45] Similarly, according to *Exod. Rab.* 2.5, when a gentile asked why God saw fit to speak to Moses from a thornbush, R. Joshua b. Qorḥah asked him whether he would have had a similar question had Moses been addressed from a carob or a sycomore. The pre-Islamic Arabian goddess al-ʿUzzā had a sanctuary containing one or more sacred trees (reportedly acacias) in a wadi near Mecca called *Suqām*.[46] At Palmyra, there was a sacred cypress in the temple of Aglibol and Malakbel called גנתא אלים "garden of the gods" in Aramaic and ἱερον ἄλσος "the sacred grove" in Greek.[47] All of this leads us to conjecture that there was a sacred sycomore in the temple renovated by the king of Qata-

[45] This text is problematic, since it names four locations, not three. In any event, the sycomore at Carmel may have been in Sycaminopolis; see chapter 1 above, and Kraeling, "Place Names," 200. For a sanctuary and a sacred grove on Mt. Carmel, see H. O. Thompson, "Carmel, Mount," in *ABD* 1. 874-75.

[46] Hišām Ibn al-Kalbī, *Kitāb al-aṣnām* (ed. A. Zeki; Cairo: Dār al-Kutub, 1924) 19 line 10; 24 line 2; 25 lines 6-11 = *The Book of Idols* (trans. N. A. Faris; Princeton: Princeton University Press, 1952) 17, 21-22; Yāqūt ibn ʿAbd Allāh, *Jacut's Geographisches Wörterbuch* (ed. F. Wüstenfeld; Leipzig: F. A. Brockhaus, 1866-73) 3. 100; J. Wellhausen, *Reste arabischen Heidentums* (Berlin: W. de Gruyter, 1961) 34, 38-39; H. M. Al-Tawil, *Early Arab Icons: Literary and Archaeological Evidence for the Cult of Religious Images in Pre-Islamic Arabia* (Ph.D. diss., University of Iowa, 1993) 138-40. Did Suqām (like Sycaminopolis and el-Jummeizeh in Palestine and Nht in Egypt, discussed in the preceding footnote and in chapter 1 n. 74 above) get its name from sycomores (*suqam*) growing there? The sanctuary was "on the road from Mecca to al-Ṭāʾif" (M. C. A. Macdonald and L. Nehmé, "Al-ʿUzzā," in *Encyclopaedia of Islam* [Leiden: E. J. Brill, 1954-] 10. 968). There are still sycomores in that area today. In a striking coincidence, Collenette (*Flowers*, 370) shows photographs of *Ficus sycomorus* taken in "Wadi Zaymah, between Taif and Makkah, on the eastern road"! According to Ibn al-Kalbī (*Aṣnām*, 18 line 8; 19 line 10 = *Idols* 16, 17), Suqām was a branch or side ravine of Wadi Ḥurāḍ in Naḫlat al-Šaʾmiyya, and these other names can also be connected with names of trees. Naḫlat al-Šaʾmiyya, also called Naḫlah (*Aṣnām*, 24 line 2; 25 line 6 = *Idols*, 21), contains the word for "palm" (*naḫlah*) (cf. Yāqūt, *Wörterbuch*, 4. 768 lines 10-11, 17; 769 lines 3, 14). Ḥurāḍ may be derived from *ḥurḍ* or *ḥuruḍ*, the name of a large shade-tree from which potash is obtained (Lane, *Lexicon*, 548). Note that *Suqām* and *Ḥurāḍ* exhibit the same vowel pattern. Note also the family name *al-Suqami* (*al-Suqāmī*?), borne by one of the hijackers who destroyed the North Tower of the World Trade Center.

[47] J. Teixidor, *The Pantheon of Palmyra* (Leiden: E. J. Brill, 1979) 36-38; idem, *The Pagan God* (Princeton, N.J.: Princeton University Press, 1977) 120-21.

ban. In view of the close connections between Yemen and Ethiopia, it is worth noting that the sycomore is "a sacred tree for various communities" in Ethiopia to this day.[48] If the same is true in Sudan, it may be possible to build a case that this cult was distributed in a continuous band from Egypt to Yemen in antiquity.[49]

From a geographic point of view, the data presented above point to Yemen and Israel.[50] We must exclude Syria, since, according to Nöldeke, "dies Wort ist den Syrern fremd."[51] Presumably, Nöldeke meant that the word שקמא in the Peshiṭta, used only as a translation equivalent of Hebrew שקמה, is a Hebraism. Indeed, it is far from clear that sycomores were found in areas where Syriac was spoken.[52]

בלס and שקמה: Lexical and Botanical Imports from South Arabia

The similarity between the distribution of the word *bls* and that of the word *šqmt*—Israel and Yemen but not Saudi Arabia, Oman or Egypt where the sycomore and other members of the genus *Ficus*

48 A. Bekele-Tesemma, *Useful Trees and Shrubs for Ethiopia* (n.p.: Regional Soil Conservation Unit, Swedish International Development Authority, 1993) 250.

49 It may have been even more widespread than that. The cult of the sycomore in modern Burundi has been compared to that in Egypt; see J. M. M. Van der Burgt, *Dictionnaire Français-Kirundi* (Bois-le-Duc, Holland: Société "L'Illustration Catholique," 1903) 556.

50 And to Ethiopia if Geez *saglā* belongs here as well. Algerian *saqūm* is uncertain. It appears, transcribed as *el seḵoum*, in a list of 49 plants capable of sustaining human life in the wilderness; M. J. E. Daumas, *La vie arabe et la société musulmane* (Paris: M. Levy frères, 1869) 381. The list was dictated in 1846 by an Algerian courier to the Arab secretary of General Eugène Daumas. Daumas identified this plant with the *Ficus sycomorus*, but it appears that he did this without the help of the informant, perhaps using a dictionary.

51 Personal communication from T. Nöldeke to I. Löw; see Löw, *Aramäische Pflanzennamen*, 386 n. 2 and *Flora*, 1. 274. The basis for Nöldeke's conclusion is not given.

52 Some modern writers list Syria among the countries in which sycomores grow, but they are generally vague about the exact locations. It is possible that these writers have drawn from older works in which Syria includes Lebanon. G. E. Post, *Flora of Syria, Palestine and Sinai* (2nd ed. by J. E. Dinsmore; Beirut: American Press, 1932-33) 516 lists locations for *F. sycomorus* in Lebanon, Palestine, and Sinai but not in Syria. I am indebted to J. Huehnergard for this reference.

flourish[53] —is surprising if the words were survivals from Proto-West-Semitic, but quite natural if the biblical sycomore was imported from Yemen together with these two lexical items.[54] The fact that the word *bls* was applied solely to the fruit in Hebrew[55] may indicate that **bls*^{3} referred primarily to fruit in ESA.[56] Alternatively, it may simply mean that, when the sycomore tree and its fruit were mentioned together in ESA, **bls*^{3}(*m*) was used only for the fruit and *s*^{1}*qmt*(*m*) was used for the tree.[57]

As we have seen, the idea that the biblical sycomore was an import from the south is far from new. It has been assumed that Egypt was the source of the import.[58] However, the linguistic data do not support

[53] For the distribution of this family in the Arabian Peninsula, see Collenette, *Flowers*, 368-70; Heller and Heyn, *Conspectus*, 28 and map 1; Miller and Cope, *Flora*, 94-102, 506-7.

[54] This suggestion is not entirely new. P. de Lagarde ("Ueber die semitischen Namen des Feigenbaums und der Feige," *Mittheilungen* [Goettingen: Dieterichsche Sortiments-buchhandlung, 1884] 1. 68) writes: "so scheint mir zweitens die Gleichung س = ס zu erweisen, daß בלס kein einheimisch israelitisches Wort ist." So too Harper (*Amos and Hosea*, 174): "the vb. seems to be a loan-word, being a denominative from the Arabic *balasun*, *a fig*, or Ethiopic *balasa* (sic) = *fig*, or *sycamore*. . . ." Of course, the correspondence cited by Lagarde is now known to be completely regular and thus cannot serve as the basis for such a suspicion.

[55] As noted in chapter 2 above, בלס still refers to the sycomore fig or some other sort of inferior fig in Mishnaic Hebrew.

[56] In Ethiopian Semitic, *balas* also refers to the tree; see chapter 2 above at n. 1. So too *balasah* in Yemeni Arabic; see Piamenta, *Dictionary* 38. However, several dictionaries of classical Arabic speak of it only as a fruit; see *Lisān al-ʿArab*, cited above; *Al-Qamūs*, cited in Bochart, *Hierozoicon*, 1. 384 (ed. Rosenmüller, 1. 406): *ʾal-balasu ṯamarun ka-t-tīni wa-t-tīnu nafsuhu* "*balas* is a fruit like the (common) fig and the (common) fig itself"; Al-Muḥkam, cited in Lane, *Lexicon*, 325 s.v. *tīn*: "*Tīn*—the tree of the *balas* or the *balas* itself." Cf. also the quotation from Baur in chapter 2 above at n. 64.

[57] The latter explanation is suggested by the Ethiopian (Geez) rendering of Jer 8:13 ואין תאנים בתאנה (LXX καὶ οὐκ ἔστιν σῦκα ἐν ταῖς συκαῖς) "and there are no figs on the fig tree": *ʾalbo balas westa saglā*. The translator has used *balas* to refer to the fruit of the fig tree, and *saglā* to refer to the fig tree itself, even though *saglā* is normally used of the sycomore tree; see Dillmann, *Lexicon*, 487. Despite the usage of his Vorlage, he apparently found it awkward to use the same word (or related words) for the tree and the fruit in the same sentence. Contrast the Jewish Yemenite expression recorded by Piamenta (*Dictionary*, 38): *taqūl mā fī l-balasah balas* "as if there are no figs on the fig tree, i. e., as if nothing has happened." Piamenta does not mention that this expression is based on Jer 8:13.

[58] See Galil, "Ancient Technique," 178, 188; Baum, *Arbres*, 21; M. Zohary, "שקמים,"

that assumption. The sycomore fig is called *k3w* (Middle Kingdom) or *k3y.w* (New Kingdom) in Egyptian; the notched (i.e., gashed) sycomore fig is called *nqʿw.t* (Dem. *3lqw*, *lqʿ*, Copt. *elkō*, *lkou*); and the sycomore tree is called *nh.t* (Dem. *nhy*, Copt. *nouhe*).[59] None of these terms bears any resemblance to Hebrew שקמה or לבס/בלס.[60]

An Egyptian export is unlikely for other reasons as well. R. K. Ritner writes: "Egyptian records often mention the importation of trees, but the reverse is unknown to me and cannot have been common."[61] The sycomore, a sacred tree and an important source of wood (not to mention food and shade) for the Egyptians, was a vital national asset, whose use was carefully controlled by the government:

> Des comtes nous renseignent sur son exploitation, qui était réglementée: le vizir des 18e et 19e dynasties était chargé de faire procéder à l'abattage des sycomores suivant les recommandations du palais.[62]

---

in אנציקלופדיה מקראית (Jerusalem: Bialik, 1972-88) 8. 258, (but Zohary rejects this view elsewhere; see below). Naturally, this assumption does not imply that the sycomore is native to Egypt; see below. For a nineteenth-century view of the Egyptian connection, see Solms-Laubach, *Herkunft*, 103: "Die Syrer dürften ihre Cultur erst von den Egyptern erlernt haben. . . ."

59 A. Erman and H. Grapow, *Wörterbuch der aegyptischen Sprache* (Leipzig: J. C. Hinrichs, 1926-63) 5. 96; 2. 343; 2. 282; W. Erichsen, *Demotisches Glossar* (Copenhagen: E. Munksgaard, 1954) 8, 264; 221; W. E. Crum, *Coptic Dictionary* (Oxford: Oxford University Press, 1939) 54b; 242b. The word for "fig" of any variety is *d3b*; it may be related to the Aramaic word תוב "wild fig," discussed in chapter 1 n. 50 above.

60 An Egyptian botanical term that does resemble לבס is *nbs*, since Egyptian *n* sometimes renders Semitic [l]; see J. E. Hoch, *Semitic Words in Egyptian Texts of the New Kingdom and Third Intermediate Period* (Princeton, N.J.: Princeton University Press, 1994) 432, 435. The *nbs*-tree is identified with the sycomore by *HALAT* s.v. בלס and J. D. W. Watts, *Vision and Prophecy in Amos* (expanded anniversary ed.; Macon, Ga.: Mercer University Press, 1997) 37 n. 54. Unfortunately, this identification is unknown to Egyptologists: "All modern authorities (ever since Maspero) are agreed that the nbs-tree is to be identified with *Zizyphus spina-christi (L.) Wild.*, more popularly known as 'Christ's thorn' in English. Germer states that this identification is certain because examples of 'Christ's-thorn' fruits have been found in Old Kingdom-period pots labelled *nbs*" (personal communication from T. Dousa). See R. Germer, *Flora des pharaonischen Ägypten* (Mainz am Rhein: P. von Zabern, 1985) 114-15. There is also a reference to a *labas*-tree in a Coptic-Arabic word list (Crum, *Coptic Dictionary*, 137b), but this sole attestation is late (c. fourteenth century C.E.) and Crum identifies it with the aloe.

61 E-mail communication from R. K. Ritner, Dec. 20, 1999.

62 Baum, *Arbres*, 23-24. In Mesopotamia, all timber cutting was regulated by the royal authorities; see P. I. Kuniholm, "Wood," in *The Oxford Encyclopedia of Archae-*

It is dangerous to extrapolate, but an Ottoman analogy is illuminating. Tristram reports that "with the Turks, as in the time of David, [the sycomore] is a royal tree, and the government claim rent for the produce wherever it is planted."[63] Finally, we should note that the sycomore is not an easy tree to smuggle out of Egypt, since it does not produce seeds there.[64]

The linguistic evidence that we have examined points not to Egypt but to Yemen. Yemen is a logical source, since it (possibly together with neighboring regions) is the only place in the world outside of Africa where the sycomore sets seeds and grows wild, thanks to the presence of its pollinating wasp.[65] In all other places where the sycomore is found, including Egypt, it is sterile and can be propagated only by means of cuttings inserted into the earth (vegetative or clonal prop-

---

ology in the Near East (ed. E. M. Meyers; New York: Oxford University Press, 1997) 5. 347. C. Müller ("Holz und Holzverarbeitung," in *Lexikon der Ägyptologie* [Wiesbaden: O. Harrassowitz, 1975-92] 2. 1265) believes that the same was true in Egypt.

[63] Tristram, *Natural History*, 398.

[64] Cf. already Theophrastus, *Enquiry*, 1. 292-93 (4.2.1): "it . . . contains absolutely no seeds." According to Galil ("Ancient Technique," 178), "dry sycomore fruit found in the grave of Ani of the XXth dynasty (about 1100 B.C.) contained neither seeds nor *Ceratosolen* wasps." For the possibility that it still reproduced spontaneously in Upper Egypt in the Predynastic period, see Baum, *Arbres*, 23, and D. Zohary and Hopf, *Domestication*, 165. For the strict border control at Egypt's eastern frontier, see M. Greenberg, *Understanding Exodus* (New York: Behrman House, 1969) 21.

[65] Precise information about other parts of the Arabian Peninsula where the sycomore grows—Asir, Hejaz, Dhofar—is difficult to obtain. Berg and Wiebes (*African Fig*, 211-12) record the presence of the pollinating wasp, *Ceratosolen arabicus* Mayr, in Yemen but nowhere else in the Arabian Peninsula. Similarly, E. Werth ("Die 'wilde' Feige im östlichen Mittelmeergebiet und die Herkunft der Feigenkultur," *Berichte der Deutschen Botanischen Gesellschaft* 50 [1932] 552; cf. also the map on p. 547, explained on p. 557) reports that the sycomore grows wild in Yemen, while "das Kulturgebiet . . . erstreckt sich . . . nach Yemen und Hedschas in Arabien." For additional information on the distribution of *Ficus sycomorus* in modern times, see Solms-Laubach, *Herkunft*, 102-3; Post, *Flora*, 516; J. Galil, M. Stein and A. Horovitz, "Origin," 191, 193; Baum, *Arbres*, 19 (and the references cited there in n. 4); C. C. Berg, "Annotated Check-list of the *Ficus* Species of the African Floristic Region, with Special Reference and a Key to the Taxa of Southern Africa," *Kirkia* 13 (1990) 256. See also the references cited in n. 26 above and in chapter 5 nn. 72-75 below. It should be noted that many descriptions overlook the distribution of *Ficus sycomorus gnaphalocarpa*; see Introduction n. 6 above. For seed production in *Ficus sycomorus*, see J. Galil and D. Eisikowitch, "On the Pollination Ecology of *Ficus Sycomorus* in East Africa," *Ecology* 49 (1968) 259-69 and idem, "Further Studies."

agation). Indeed, Schweinfurth, who saw wild sycomores in Yemen, posited a Yemeni origin for the Egyptian sycomore tree itself.[66]

The sycomore would not be the only tree in the Levant thought to have been imported from South Arabia. A tree that grows today both in Lebanon and in South Arabia is believed by some to be the biblical *almog*, transplanted in Solomon's time (1 Kgs 10:11-12).[67] A number of botanists have suggested that the carob (which, as noted above, was closely associated with the sycomore in rabbinic and classical literature)[68] was brought to Palestine from Yemen in antiquity.[69] The same goes for *Commiphora opobalsamum*. This tree grew in royal groves near Jericho, according to Theophrastus, Strabo, Pliny, and Josephus, but it is indigenous to South Arabia,[70] and may have been transplanted from there.[71] It not surprising, then, that Josephus records a tradition connecting this tree with the Queen of Sheba: ". . . and they say that we still have the root of the opobalsamon, which our country still bears, as result of this woman's gift."[72]

[66] Schweinfurth, "Kulturgeschichte," 34-35; cf. idem, "Sitzungs-Bericht," 158. He was followed by Henslow, "Egyptian Figs"; idem, "The Sycomore Fig," 130. Cf. what Condit (*Fig*, 9) writes about the common fig: "The fig tree was probably first cultivated in the fertile part of southern Arabia, where wild specimins (sic), such as those reported in 1923 by C. M. Doughty, are still found."

[67] M. Elat, קשרי כלכלה בין ארצות המקרא בימי בית ראשון (Jerusalem: Bialik, 1977) 60.

[68] See chapter 2 nn. 33-34 above.

[69] M. Kislev, ההסטוריה של החרוב בארץ, *Halamish* 6 (1988) 25-27. Based on the assumption that the carob is not mentioned in the Bible, Kislev entertained the possibility that the carob was brought from Yemen in postbiblical times. It is true that the word חרוב does not appear in the Bible, but it has been shown that חרי יונים in 2 Kgs 6:25 is a synonym, perhaps colloquial, of חרוב; see M. Held, "Studies in Comparative Semitic Lexicography," in *Studies in Honor of Benno Landsberger on his Seventy-fifth Birthday April 21, 1965* (AS 16; Chicago: The University of Chicago Press, 1965) 395-98 and the literature cited there. Moreover, any attempt to show that the carob was imported to Palestine in postbiblical times would seem to be refuted by the archeobotanical evidence (pollen grains, seeds and charred wood of carob trees) cited in N. Liphschitz, "*Ceratonia Siliqua* in Israel: An Ancient Element or a Newcomer," *Israel Journal of Botany* 36 (1987) 194-95. I am indebted to M. Kislev for this reference.

[70] See R. C. Steiner, *The Case for Fricative-Laterals in Proto-Semitic* (AOS 59; New Haven: American Oriental Society, 1977) 126-27, 129 and add Josephus, *B.J.* 1.6.6 §138; 1.18.5 §361; 4.8.3 §469; *A.J.* 14.4.1 §54; 15.4.2 §96.

[71] For other possibilites, see A. C. Western, "The Ecological Interpretation of Ancient Charcoals from Jericho," *Levant* 3 (1971) 37.

[72] *A.J.* 8.6.6 §174.

The sycomore could have been brought to Palestine from Yemen by traders. Israel's commercial ties with the Kingdom of Sheba in Yemen in the time of Solomon (tenth century B.C.E.) are well known.[73] Judging from the biblical account, Solomon imported a great deal of wood: cedar and juniper from Tyre (1 Kgs 5:22, 24) and *almog* wood transported by Hiram from Ophir (1 Kgs 10:11-12). As noted above, it has been argued that Solomon's *almog* wood came from saplings transplanted from South Arabia to Lebanon.[74]

The Arabian trade did not begin with Solomon. L. E. Stager writes: "In Sheba grew the best aromatics in the world. . . . By the Late Bronze Age, the aromatics trade had become the most lucrative business in the ancient Near East thanks to the dromedary camel."[75] I. Finkelstein shows that overland trade routes from Arabia to the southern Shephelah and the coastal plain were active already by the twelfth century B.C.E.[76]

It seems likely, then, that sycomore figs and/or saplings were brought to Israel from Yemen at some point during the two centuries preceding Solomon's reign[77] and that the words for the fig (*bls*) and the tree (*šqmt*) were brought with them.[78] M. Zohary's claim that "there is

[73] For a recent discussion of these ties, see L. E. Stager, "Forging an Identity: The Emergence of Ancient Israel," in *The Oxford History of the Biblical World* (ed. M. D. Coogan; New York-Oxford: Oxford University Press, 1998) 145-46; see also I. Finkelstein, "Arabian Trade and Socio-Political Conditions in the Negev in the Twelfth-Eleventh Centuries B.C.E.," *JNES* 47 (1988) 251; J. S. Holladay, Jr., "The Kingdoms of Israel and Judah: Political and Economic Centralization in the Iron IIA-B (ca. 1000-750 BCE)," in *The Archaeology of Society in the Holy Land* (ed. T. E. Levy; London: Leicester University Press, 1998) 383-86; Jaruzelska, *Amos*, 94-99.

[74] Elat, קשרי כלכלה, 60.

[75] Stager, "Forging," 146.

[76] Finkelstein, "Arabian Trade," 247-48.

[77] Cf. no. 1 above.

[78] It is tempting to appeal to the foreign origin of this word to account for its non-segolate plural in Hebrew (viz., שִׁקְמִים instead of שְׁקָמִים*), but other botanical terms exhibit the same irregularity: בָּטְנִים, הָבְנִים, פִּשְׁתִּים, and צֶאֱלִים; see A. Schlesinger, כתבי עקיבא שליזנגר; מחקרים במקרא ובלשונו (Jerusalem: Israel Society for Biblical Research, 1962) 50-51; J. Blau, "Marginalia Semitica I," *Israel Oriental Studies* 1 (1971) 7 = J. Blau, *Topics in Hebrew and Semitic Linguistics* (Jerusalem: Magnes, 1998) 191. At first glance, the Greek form συκάμιν-, with *alpha* inserted between the second and third consonants, seems to exhibit the regular segolate plural; but see at n. 86 below.

no evidence that [the sycomore] was imported into this country"[79] is no longer tenable.

The Etymology of שקמה

Our conclusion that שקמה is a South Arabian loanword attested in Qatabanian has relevance for a rather speculative etymology proposed by P. Haupt:

> The Hebrew name of the sycamore trees, *šiqmîm*, . . . may be an old causative (AJSL 23, 248) derived from the root *qm*; the original meaning may be *staturosa*; cf. *gĕḇah qômâ*, lofty of stature, Ezek. 31:3. The *ficus Ægyptia* may reach a height of 50 feet.[80]

Haupt assumes that Biblical Hebrew preserves relics of an old *š*-causative. If this view presupposes direct preservation of *š*-stem forms from Proto-Semitic, many scholars may find this assumption questionable.[81] However, the relic could have been preserved indirectly, via borrowing from a Semitic language in which the *š*-causative was the norm. Qatabanian happens to have been a Semitic language of this type.[82] Thus, the inscription quoted above has the causative s^1*ḥd*[*ṯ*] "renewed" five words before s^1*qmtm*.[83]

Ezek 31:3 is not the only verse in which קומה refers to the stature of a tree. The same is true of קומה in 2 Kgs 19:23 = Isa 37:24, Ezek 17:6, and Cant 7:8. The situation in South Arabian is similar. The verb *qwm* appears in a Sabaic inscription (Gl 1520/4) with the name of a tree (*ʿlb*

[79] M. Zohary, *Plants*, 68.

[80] P. Haupt, "Was Amos a Sheepman?" *JBL* 35 (1916) 282.

[81] Even for Biblical Aramaic, the tendency of most scholars has been to view examples of the *š*-causative as loanwords; see C. Rabin, "The Nature and Origin of the Šafʿel in Hebrew and Aramaic," *ErIsr* 9 (1969) 148-58; Kaufman, *Influences*, 123-24; E. Y. Kutscher, "Aramaic," in *Current Trends in Linguistics* (ed. T. A. Sebeok; 14 vols.; The Hague: Mouton, 1963-) 6. 354.

[82] See A. F. L. Beeston, *Sabaic Grammar* (JSS Monograph 6; Manchester: JSS, 1984) 64.

[83] The single example of the causative of *qwm* cited by Ricks (*Lexicon*, 144) has an *h*-preformative: *hqmhw* "he set it up." However, J. Huehnergard (personal communication) notes that the object suffix in *h* shows that it is Sabaic. Cf. F. Bron, "Le bilinguisme en Arabie du Sud préislamique," in *Mosaïque de langues, mosaïque culturelle: le bilinguisme dans le Proche-Orient ancien* (ed. F. Briquel-Chatonnet; Paris: Maisonneuve, 1996) 125-30.

"Zizyphus spina-Christi") as its subject: *ʾl yqwm kl ʿlbm bfnwtn* "werde nicht angelegt jegliche *ʿlb*-Pflanzung an d(ies)em Kanal."[84] Since the meaning of *qwm* in this passage is "stand, be planted,"[85] its causative would have to mean "make stand, plant." One might then claim that the שקמה gets its name from the fact that it causes buildings to stand; however, the use of an inanimate noun as the subject of a causative verb would be anomalous. A more grammatical solution would be to assume that the tree gets its name from the fact that it is caused to stand, i.e., planted. Thus, if the etymon of שקמה is a causative of a verb meaning "stand," it is probably also a passive; cf. the German expression *mit Bäumen bestanden* "planted with trees." Such a form could have been vocalized something like *šūqamat* (cf. Hebrew הוּקְמָה, הוּקַם) or *šuqamat*.[86] That is the vocalization that seems to be reflected in Christian Palestinian Aramaic שוקמא, Greek συκάμινος, and perhaps Arabic *sawqam* as well.[87] The name could allude to a belief that the sycomore was planted by the gods, possibly the sycomore-gods discussed earlier in this chapter. It is even possible that the Qatabanian term *s¹qmtm* is an abridgment of a phrase like **ms¹qmt ʾlhn* "planted by the gods."

In sum, our conclusion that שקמה is a South Arabian loanword eliminates some of the objections to Haupt's etymology; nevertheless, in the absence of further evidence, it must remain nothing more than an intriguing possibility.

[84] M. Höfner and J. M. Solá Solé, *Inschriften aus dem Gebiet zwischen Mārib und dem Ġōf* (Sammlung Eduard Glaser 2; Vienna: H. Böhlhaus, 1961) 19-20.

[85] A. F. L. Beeston, M. A. Ghul, W. W. Müller, and J. Ryckmans, *Sabaic Dictionary* (Louvain-la-Neuve: Peeters, 1982) 110 s.v.

[86] The vocalization of the corresponding ESA form is unknown; see Beeston, *Sabaic Grammar* 14 §5.3. J. Huehnergard suggests (personal communication) that the Hebrew vocalization may be the result of repatterning on the basis of other Hebrew botanical terms, such as those in n. 78 above. He adds that such repatterning is common in Arabic, as demonstrated by J. T. Fox, *Semitic Noun Patterns* (HSS 52; Winona Lake, Ind.: Eisenbrauns, 2003).

[87] See n. 29 and chapter 1 above.

CHAPTER 4

בוקר, בנקדים, and מאחרי הצאן

The Meaning of בוקר

The participle בוקר (Amos 7:14), also a *hapax legomenon*, was recognized as a denominative long before בולס was. Ibn Janāḥ, for example, says explicitly that it is derived from the word בקר "cattle": "and from [בקר] it says 'I am a בוקר,' i.e., an owner of cattle not in need of anyone."[1] Even earlier, Jerome reports that Aquila, Symmachus and Theodotion made a point of rendering *boger* (*sic,* instead of the expected *boker*) with βούκολος, which refers to "one who pastures cattle, not sheep."[2]

Most of the medieval exegetes follow the Targum and the Talmud (*b. Ned.* 38a) in taking the בוקר to be something more than a simple cowherd.[3] According to these commentators, Amos identifies himself as a בוקר in order to rebut the condescending insinuation that he has come to Bethel looking for a handout; hence, the בוקר must be reason-

[1] Ibn Janāḥ, *ʾUṣūl*, 106 lines 9-10.

[2] Jerome, *Commentarii*, 324 lines 387-88.

[3] So Yefet (Ms. British Library Or. 2400 = Margoliouth 282, p. רב = f. 102b lines 12-13), Al-Fāsī (*Jāmiʿ al-Alfāẓ* 1. 265 lines 84-87), Ibn Janāḥ (*ʾUṣūl*, 106 lines 9-10 and 451 lines 9-10), and Ibn Ezra (פירושי תרי־עשר, 310-11; less clearly 248-49). Cf. n. 58 below.

ably high on the socioeconomic scale. This deduction makes good sense. A person hired to herd cattle was probably called a רועה בקר "cowherd," as in the Mishnah (*m. Sanh.* 3.1), just as a person hired to herd sheep and goats was called a רועה צאן "shepherd." The term בוקר probably denoted a cattleman, a man who bred and sold cattle.

In modern times, however, several objections have been raised to this interpretation of בוקר as a denominative of בקר. They are summarized by Y. Breslavy: (1) the interpretation creates a contradiction between 7:14 (where Amos describes himself as a בוקר) and 1:1 (where Amos is said to be בנקדים);[4] (2) the interpretation creates a contradiction between 7:14 (where Amos calls himself a בוקר) and 7:15 (where Amos says he was taken מאחרי הצאן);[5] (3) the interpretation calls for a different nominal pattern, viz., בַּקָּר as in *y. Beṣah* 5.3, 63b; cf. Arabic *baqqār* and Mishnaic Hebrew חַמָּר, גַּמָּל;[6] (4) "The mountain region, and the Judean mountains in particular, are not good for raising cattle, and certainly Tekoa, . . . situated on the threshold of the desert, was not fit for raising cattle. This border region can be exploited only by sheep and goats, since only they are capable of climbing the slopes of the mountains and hills and sustaining themselves for most of the year from poor dry grasses and a large series of desert plants from which cattle cannot derive benefit in any way."[7]

These objections, especially (1) and (2), have inspired a wide variety of ingenious reinterpretations and emendations[8] connecting the participle בוקר with

[4] Y. Breslavy, עמוס—נוקד, בוקר ובולס שקמים, *Bet Mikra* 31 (1966-67) 93.

[5] Breslavy, עמוס, 92.

[6] Breslavy, עמוס, 94. Cf. also Late Aramaic בַּקָּר.

[7] A. Cohen *apud* Breslavy, עמוס, 92; cf. Weiss, ספר עמוס, 2. 445 n. 161.

[8] See the sources cited in Weiss, ספר עמוס, 2. 445 nn. 162-68, especially J. Wright, "Did Amos Inspect Livers," *AusBR* 23 (1975) 6-11. The earliest reinterpretation is perhaps that which interprets בוקר as "shepherd," comparing the *piʿel* of this root in Ezek 34:11-12: ודרשתי את צאני ובקרתים: כבקרת רעה עדרו . . . אבקר את צאני "I shall search for my flock and seek them out; as a shepherd seeks out his flock . . . so shall I seek out my flock." A less common approach retains the traditional understanding of בוקר, while minimizing the value of אשר היה בנקדים in the superscription and reinterpreting or deleting מאחרי הצאן; see H. Schult, "Amos 7_{15a} und die Legitimation des Aussenseiters," *Probleme biblischer Theologie: Gerhard von Rad zum 70. Geburtstag* . . . (ed. H. W. Wolff; Munich: C. Kaiser, 1971) 463-74 and O. Loretz, "Die Berufung des Propheten Amos (7,14-15)," *UF* 6 (1974) 487-88.

1. sheep
   "one who looks after"
   "shepherd" (l. נוקד)
2. sycomore figs
   "one who looks after"
   "splitter"
   "examiner"
   "gleaner" (l. בוצר)
   "puncturer, piercer" (l. נוקד or דוקר)
3. prophecy
   "seer"
4. hepatoscopy
   "examiner."

In addition to these non-denominative interpretations, there is also a reinterpretation that takes בוקר as an Aramaism, a denominative of Aram. בקרא, with the latter taken to mean "Herde im allgemein" based on Syriac usage.[9]

Many of these solutions create new problems. A participle meaning "one who looks after," "splitter," or "puncturer, piercer" is not sufficiently specific to be the name of an occupation. An objective genitive would be required to remedy this defect, e.g., מבקר צאן or בוקר/דוקר שקמים. Some have suggested that the word שקמים later in the verse functions as the object of בוקר as well as בולס, but this suggestion presupposes a number of linguistic anomalies. First, there is the problem of the intervening pronoun, אנכי, noted by Weiss.[10] This problem is, in my view, insurmountable, because a phrase like בוקר/דוקר שקמים would normally be a genitive construction, if it is the name of an occupation, parallel to נביא, rather than merely the description of an activity.[11] However, the position of אנכי makes it impossible for בוקר and

[9] B. J. Diebner, "Berufe und Berufung des Amos (Am 1,1 und 7,14f.)," *Dielheimer Blätter zum Alten Testament und seiner Rezeption in der Alten Kirche* 23 (1986) 103-4, 119. Cf. C. F. Keil and F. Delitzsch, *The Twelve Minor Prophets* (trans. J. Martin; Edinburgh: T. & T. Clark, 1868) 312. Cf. also Mandaic באקרא "herd, flock."

[10] Weiss, ספר עמוס, 2. 448 n. 195. Weiss cites Rudolph, *Joel—Amos—Obadja—Jona*, but I find no discussion of this issue there.

[11] Contrast והאנשים רעי צאן "and the men are shepherds" (Gen 46:32) with כל ימי היותנו עמם רעים הצאן "the entire time that we were with them herding the צאן" (1 Sam 25:16).

שקמים to form a genitive construction, at least in prose.[12] And even if we were to move the pronoun, we would still be left with the ungrammatical בוקר/דוקר ובולס שקמים.[13] In addition, some of the interpretations require the *piʿel* form (מבקר) rather than the *qal* (בוקר). As for the claim that בוקר is an Aramaism, it rests on the assumption that Aramaic בקרא could be used of sheep and goats in the biblical period, which cannot be proven using one or two Late Aramaic dialects; indeed, it is not even clear that the word was in use in Aramaic in that period, since it is not attested before Late Aramaic. In view of these flaws, we must conclude that, from a purely linguistic point of view, the traditional interpretation remains superior to the others.

As for the objections raised against the traditional interpretation, they are easily answered. The first two objections will be answered in the remainder of this chapter with the following claims: Amos was an איש מקנה (cf. Gen 46:32, 34), who owned both צאן (7:15) and בקר (7:14).[14] More precisely, he was a נוקד in the broad sense of the Neo-Babylonian *nāqidu ša ṣēni u lâti*, a specialist stockbreeder who owned part of the herds and flocks that he managed. He alludes to his בקר as a sign of self-sufficiency[15] and to his צאן as a symbol of legitimacy.[16]

The third objection seems somewhat frivolous. The participle too is used for professions, especially in this semantic field, e.g., רועה, נוקד. Nor does the fact that this is a denominative change matters. H. Weippert correctly compares בוקר with כורם,[17] just as Bochart compared בולס with כורם. All are denominative participles used to refer to professions.

[12] See, for example, P. Joüon and T. Muraoka, *A Grammar of Biblical Hebrew* (Subsidia Biblica 14; Rome: Pontifical Biblical Institute, 1991) 2. 463-65.

[13] Hebrew grammar requires בוקר/דוקר שקמים ובולסם or the like in prose; see Joüon and Muraoka, *Grammar*, 464.

[14] So H. Schmidt, *Der Prophet Amos* (Tübingen: J. C. B. Mohr, 1917) 9 n. 1; H. J. Stoebe, "Der Prophet Amos und sein bürgerlicher Beruf," *Wort und Dienst: Jahrbuch der Theologischen Schule Bethel* 51 (1957) 177; O. Eissfeldt, *The Old Testament: An Introduction* (Oxford: B. Blackwell, 1965) 396 n. 5; Rudolph, *Joel—Amos—Obadja—Jona*, 114; S. M. Paul, *Amos* (Hermeneia; Minneapolis: Fortress, 1991) 35, 247-48.

[15] They were roughly ten times more valuable than צאן; see E. Firmage, "Zoology," in *ABD* 6. 1119b.

[16] See below for a more precise formulation.

[17] H. Weippert, "Amos: Seine Bilder und ihr Milieu," in *Beiträge zur prophetischen Bildsprache in Israel und Assyrien* (ed. H. Weippert, K. Seybold, and M. Weippert; Freiburg: Universitätsverlag, 1985) 4 n. 8.

The fourth objection appears to be somewhat exaggerated. If the wide plain around Tekoa is suitable for farming,[18] it is certainly suitable for grazing cattle. As for Tekoa itself:

> The hill, four or five acres, is broad at the top and not steep. The country is sterile and rocky, but rich in pasturage.[19]

H. J. Stoebe reports encountering a bedouin woman tending sheep and a cow while climbing from Herodion to Tekoa in the fall of 1955.[20] According to the 1974 animal census of the West Bank, cattle are raised in the hill country and even in the "desert fringe."[21] As for ancient Israel, cattle bones are found in archeological excavations in the hill country and in desert areas.[22] The proportion of cattle to sheep and goats is lower today than in biblical times, "no doubt due to the fact that mechanized farming has largely replaced the use of the ox-drawn plow."[23]

The fourth objection is also irrelevant or, at least, inconsistent. The conditions that make Tekoa inhospitable to bovines are lethal to sycomores. As a result, most scholars, including Breslavy himself, look for Amos' sycomores outside of Tekoa.[24] Why not do the same for Amos' cattle? Indeed, as we shall argue below, there are good reasons for assuming that Amos' animals spent at least part of the year in the vicinity of his sycomores.

The Meaning of בנקדים

The scholarly debate about the term נוקד (Amos 1:1 בנקדים, 2 Kgs 3:4 נקד) has focused on three questions: (1) Is the term restricted to people who deal with צאן or is it also used of people who deal with בקר?[25]

[18] As reported by Z. Kallai, "Tekoa," in אנציקלופדיה מקראית (Jerusalem: Bialik, 1972-88) 8. 925.

[19] Harper, *Amos and Hosea*, 3.

[20] Stoebe, "Der Prophet Amos," 177.

[21] Firmage, "Zoology," 1120b.

[22] Firmage, "Zoology," 1120a.

[23] Firmage, "Zoology," 1120b.

[24] See chapter 5 below.

[25] Throughout this monograph, I use the Hebrew terms צאן and בקר, due to the difficulty of finding an English translation for צאן. All of the translations currently in

(2) Does the term have socioeconomic connotations? (3) Does the term have sacral overtones?

The first question, unlike the second and third, has bothered exegetes for a long time. Abarbanel's uncertainty about the meaning of נוקד is evident in the corrections made in the autograph of his commentary: (1) רועה צאן בנוקדים אשר בעיר תקוע "a herder of צאן among the נוקדים in the city of Tekoa" → רועה בנוקדים אשר בעיר תקוע "a herder among the נוקדים in the city of Tekoa"; (2) נוקד הוא בעל הצאן "a נוקד is an owner of צאן" → נוקד הוא בעל הצאן או הבקר "a נוקד is an owner of צאן or בקר"; (3) אבל אמר אשר היה בנוקדים להגיד שהיה רועה צאן אדם אחר "but it says that he was 'among the נוקדים' to indicate that he herded צאן belonging to another man" → אבל אמר אשר היה בנוקדים להגיד שהיה רועה מקנה אדם אחר "but it says that he was 'among the נוקדים' to indicate that he herded livestock belonging to another man."[26] The ancient versions do not speak with one voice concerning this question. For Aquila, the נוקד is a ποιμνιοτρόφος, a rearer of צאן,[27] both in 2 Kgs 3:4 and in Amos 1:1.[28] Similarly, for Symmachus (according to one tradition), he is a τρέφων βοσκήματα, a rearer of צאן[29] (2 Kgs 3:4), or a ποιμήν, a shepherd (Amos 1:1).[30] Other translators, however, used a more general term at Amos 1:1: κτηνοτρόφος "rearer of livestock."[31] The expression ἄνδρες κτηνοτρόφοι is used by the Septuagint to render אנשי מקנה in Gen 46:32, 34. The word κτῆνος regularly renders מקנה "livestock (צאן ובקר)." Thus, at Gen 26:14, κτήνη προβάτων καὶ κτήνη βοῶν translates מקנה צאן ומקנה בקר.

The Targum's rendering—מרי גיתי(ן) in 2 Kgs 3:4 and Amos 1:1—is similar to κτηνοτρόφος in meaning and usage. Targumic Aramaic גיתי is a loanword related to Avestan *gaēθa* "possession," and it is used to render מקנה "livestock."[32] Thus, at Gen 46:32 כי אנשי מקנה היו וצאנם

use—"sheep and goats," "sheep-goats," "flocks," "small cattle," "ovi-caprine cattle," "caprovids," "capridae"—have drawbacks.

[26] Abarbanel, *Comentario*, 18 line 11; 18 line 20; 18 line 21 - 19 line 1.

[27] In the usage of Aquila, ποίμνιον normally renders צאן.

[28] Origen, *Hexapla*, 1. 655, 2. 967.

[29] In the usage of Symmachus, βοσκήματα always renders צאן.

[30] Origen, *Hexapla*, 1. 655, 2. 967.

[31] Origen, *Hexapla*, 2. 967. For Philo's distinction between this term and ποιμήν, see P. de Robert, *Le berger d'Israël* (Neuchâtel: Delachaux et Niestlé, 1968) 19 n. 4. Philo's analysis does not seem to fit the usage of the translators.

[32] A. Tal, לשון התרגום לנביאים ראשונים ומעמדה בכלל ניבי הארמית (Tel-Aviv: Tel-Aviv

ארי גברי מרי גיתי הוו ועבקרם וכל אשר להם הביאו, Onqelos translates ארי גברי מרי גיתי הוו וענהון ותוריהון וכל דילהון איתיו "for they are owners of גיתי, and they have brought their צאן and their בקר and all that is theirs," showing that גיתי includes בקר as well as צאן. The same picture emerges from Onqelos to Gen 26:14: גיתי ענא וגיתי תורי. In Targum Jonathan to the Prophets, too, גיתי includes בקר. This translation uses מרי גיתי(ן) for בוקר in Amos 7:14[33] and for נוקד in 2 Kgs 3:4, where it takes כרים to mean תורין דפטמא "fattened cattle."[34] We may say that, in the view of the Aramaic translators, אנשי מקנה = (גברי) מרי גיתי = נקדים.

The current view of the נוקד as a shepherd has less to do with the versions than with the lingering influence of the "pan-Arabic" period of biblical lexicography. Hebrew נוקד was already compared with Arabic *naqqād* by Judah Ibn Quraysh (late ninth or early tenth century).[35] This Arabic term has a very restricted meaning. As noted by Ibn Janāḥ, it denotes a *rāʿī naqad*, a shepherd who tends a specific variety of short-legged sheep called *naqad*.[36]

Ibn Janāḥ was wise enough to distinguish the very narrow meaning

University, 1975) 186, 187. Despite its etymology, it is never used to render קנין "possession" or רכוש "property." Nor is it used to render צאן; hence, M. Ellenbogen's translation (*Foreign Words in the Old Testament: Their Origin and Etymology* [London: Luzac, 1962] 115) of מרי גיתי(ן) in 2 Kgs 3:4 and Amos 1:1 as "flock owner" is unacceptable. The same goes for Watts' translation (*Vision*, 35 and 36 n. 48) of מרי גיתי(ן) at Amos 1:1 and 7:14 as "sheep-master" and the conclusions that he derives therefrom. In support of his translation, Watts cites S. Speier, "Bemerkungen zu Amos," *VT* 3 (1953) 305. There is nothing relevant there on p. 305, but on p. 308, Speier writes: ". . . גיתי die aramäische Übersetzung des hebräischen מקנה, Vieh, ist. . . ."

33 Cf. the citation in *b. Ned.* 38a.

34 Distinguished from תורין דרעיא "pastured cattle" in the Targum to 1 Kgs 5:3. For the distinction in earlier times, see M. Stępień, *Animal Husbandry in the Ancient Near East: A Prosopographic Study of Third-Millennium Umma* (Bethesda, Md.: CDL, 1996) 27, 197-98. These forms are vocalized דִפְטְמָא and דִרְעְיָא by Sperber in *The Bible in Aramaic,* 2. 221 (1 Kgs 5:3) and 2. 275 (2 Kgs 3:4). And דִרְעְיָא is translated "of the pasture" in D. J. Harrington and A. J. Saldarini, *Targum Jonathan of the Former Prophets* (Wilmington, Del.: M. Glazier, 1987) 220, 268. However, the Hebrew counterparts of these terms in *b. Beṣah* 38a are שור של פטם and שור של רועה. The correct vocalization must therefore be תורין דפַטָּמָא "cattle of the fattener" and תורין דָרָעְיָא "cattle of the shepherd."

35 Judah Ibn Quraysh, הרִסַאלַה׳ של יהודה בן קוּרַיְש (ed. D. Becker; Tel-Aviv: Tel-Aviv University, 1984) 276-77.

36 Ibn Janāḥ, *ʾUṣūl*, 451 line 6. Ibn Janāḥ does not give a definition for Arabic *naqad*, but later exegetes do; see Poznański, "Ibn Balʿam," 28; Tanḥum Yerushalmi, פירוש לתרי־עשר, 70-71; Bochart, *Hierozoicon*, 1. 442-43.

of this Arabic cognate from the meaning of Hebrew נוקד (*ṣāḥib ġanam* "an owner of sheep").[37] W. Gesenius, standing on the broad shoulders of Ibn Janāḥ, widened the meaning of נוקד even further. After presenting the meaning of the Arabic cognate, he writes: "Im Hebr. war aber wohl die Bedeutung allgemeiner: Viehhirt."[38] Other Hebraists ignored Ibn Janāḥ's example. S. R. Driver asserted that the herdsmen of Tekoa, "as the word used implies, reared a special breed of sheep, of small and stunted growth, but prized on account of their wool."[39] To emphasize the point, he translated נקדים as "*nakad*-keepers."[40] T. Nöldeke was of the same opinion. When an Akkadian cognate was discovered, he dismissed its importance: "Was נֹקֵד eigentlich bedeute, erfahren wir erst durch die Erklärung von *naqad*."[41]

In 1904, J. A. Montgomery adduced the Akkadian cognate as evidence that the Arabic meaning was too narrow:

> The current interpretation of the word explains it from the Arabic *nakad*, which is defined by Freytag (*Lex. s.v.*) as "a deformed and short-legged race of sheep which abounds in the Arabian province of Bahrein, . . . whose wool is considered to be the very finest." . . .
>
> But the Arabic parallel is provincial, and it seems a far cry to use such a special term for the general designation of shepherd as applied to Amos or Mesha. Should not the word, therefore, be explained as the equivalent of the Assyrian *nâkidu*, shepherd?[42]

As Akkadian gradually supplanted Arabic as the mainstay of biblical lexicography, the answer to Montgomery's question came to be viewed as self-evident. Driver's view of the נוקד, cited above from works published in 1891 and 1897, is nowhere to be found in BDB (1907), which translates "sheep-raiser, -dealer, or -tender."[43] This shift reached its

[37] Ibn Janāḥ, *ʾUṣūl*, 451 line 5.

[38] W. Gesenius, *Hebräisches and chaldäisches Handwörterbuch über das Alte Testament* (2nd ed.; Leipzig: F. C. W. Vogel, 1823) 509 s.v.

[39] S. R. Driver, *An Introduction to the Literature of the Old Testament* (New York: Charles Scribner's Sons, 1891) 293.

[40] Driver, *Books*, 128.

[41] T. Nöldeke, Review of F. Delitzsch, *Prolegomena eines neuen hebräisch-aramäischen Wörterbuchs zum Alten Testament*, *ZDMG* 40 (1886) 723.

[42] J. A. Montgomery, "Notes on Amos," *JBL* 23 (1904) 94.

[43] BDB 667, s.v. According to the preface, Driver was not assigned this entry, but he "read all the proofs and made many suggestions" (p. ix).

natural conclusion in 1962 with a claim that Hebrew נוקד was in fact borrowed from Akkadian.[44]

Montgomery's push to broaden the interpretion of נוקד did not go far enough. It was based on an interpretation of the Akkadian cognate that was itself too narrow, an interpretation that became increasingly untenable as the study of Akkadian progressed. In a highly influential series of articles beginning in 1948, M. San Nicolò showed that the Babylonian *nāqidu* could be in charge not only of צאן (*nāqidu ša ṣēni*) but also of בקר (*nāqidu ša lâti*) or צאן ובקר (*nāqidu ša ṣēni u lâti*).[45] He noted further that *NA.GAD* = *nāqidu* is used of "the breeder of sheep, asses, and cows" already in the Sumerian Fara texts.[46] Finally, he pointed to a *nāqidu* of the Eanna temple in Uruk (555/54 B.C.E.) named Iqišā son of Nannā-ereš who was in charge of 500 cows in addition to 2000 sheep and goats.[47]

These 2,500 animals have been mentioned by many an author in connection with question (2), but, inexplicably, they have been totally ignored—sometimes by the very same author—in dealing with question (1). Thus, A. S. Kapelrud writes: "This *naqidu* . . . might be responsible for 500 cows and 2000 sheep and goats. The term . . . is

[44] Ellenbogen, *Foreign Words*, 115. Reaction to this claim has been mixed. P. V. Mankowski (*Akkadian Loanwords in Biblical Hebrew* [Winona Lake, Indiana: Eisenbrauns, 2000] 104 n. 372) is not convinced. *HALAT*, 679 s.v., agrees with Ellenbogen, but, instead of citing him, it refers to two works that disagree: H. Zimmern, *Akkadische Fremdwörter als Beweis für babylonischen Kultureinfluss* (Leipzig: J. C. Hinrichs, 1915) 41 and *AHw* 744a. The claim of the latter is that only Sumerian and Syriac got the word from Akkadian; the suggestion of the former is that Akkadian may have borrowed the word from West Semitic! See also chapter 5 n. 34 below.

[45] M. San Nicolò, "Materialien zur Viehwirtschaft in den neubabylonischen Tempeln. I," *Or* NS 17 (1948) 284; idem, "Materialien zur Viehwirtschaft in den neubabylonischen Tempeln. II," *Or* NS 18 (1949) 295. Cf. H. M. Kümmel, *Familie, Beruf und Amt im spätbabylonischen Uruk* (Berlin: Mann, 1979) 49. For the problems associated with the reading *lâtu*, see M. Van De Mieroop, "Sheep and Goat Herding According to the Old Babylonian Texts from Ur," *Bulletin on Sumerian Agriculture* 7 (1993) 173-74 and G. van Driel, "Cattle in the Neo-Babylonian Period," *Bulletin on Sumerian Agriculture* 8 (1995) 217-18.

[46] San Nicolò, "Viehwirtschaft I," 284 n. 3; cf. B. Landsberger, *Materialien zum sumerischen Lexikon, II: Die Serie Ur-e-a* = nâqu (Rome: Pontificium Institutum Biblicum, 1951) 107 n. 3.

[47] San Nicolò, "Viehwirtschaft I," 285, cf. 279-81.

found also in the Sumerian Fara texts, with the meaning shepherd."[48] Kapelrud cites San Nicolò here, but his statement about the Fara texts contradicts San Nicolò's own words, which we have cited above. In the next paragraph, Kapelrud continues: "The translation sheep-raiser [for *noqed*], as suggested by Koehler, may therefore be accepted." A. Jeffers makes the contradiction even more blatant: "[The *nāqidu*'s] were often [*sic!* RCS] responsible for five hundred cows and two thousand sheep and goats. Moreover, the term 'sheep-raiser' is very easy to derive from the original meaning of the root *nqd*, 'to puncture.' . . ."[49]

Nor has the publication of the Akkadian dictionaries succeeded in uprooting the tradition among biblicists of restricting the *nāqidu* to sheep, even though *AHw* (s.v. *nāqidum* "Hirte") cites the phrase *nāqidu ša lâti*, and *CAD* (s.v. *nāqidu* "herdsman") begins by citing Code of Hammurabi §261.22: *šumma awīlum nāqidam ana alpī u ṣēnī re'îm īgur* "if a man hires a herdsman to herd cattle or sheep and goats."[50] Segert cites these very entries in claiming:

> Larger numbers of sheep may help to explain the apparent specialization of the term *nāqidu* for herdsmen of sheep, while the term expressing rather the overseeing of animals at grazing, Akkadian *rē'û*, Ugaritic *rʿy*, Hebrew *rōʿē* could be in some contexts used as terms for herdsmen of bovines, of large cattle.[51]

Delcor too refers to *AHw* in asserting that "l'accadien *naqidu* est traduit par 'berger, éleveur de moutons.'"[52] This reading of the Akka-

[48] A. S. Kapelrud, *Central Ideas in Amos* (Oslo: Universitetsforlaget, 1961) 6.

[49] A. Jeffers, *Magic and Divination in Ancient Palestine and Syria* (Leiden: E. J. Brill, 1996) 112.

[50] *AHw*, 744a; *CAD* vol. *N*, 333.

[51] S. Segert, "The Ugaritic *nqdm* After Twenty Years. A Note on the Function of Ugaritic *nqdm*," *UF* 19 (1987) 410. Similarly, P. V. Mankowski (*Akkadian Loanwords in Biblical Hebrew* [HSS 47; Winona Lake, Indiana: Eisenbrauns, 2000] 104 n. 372) writes that connecting *nāqidu* with Arabic *naqada* ("to pick, separate the good from the bad") "would fit with the notion of *nāqidu* as *Schafhirt* as distinct from *Rinderhirt*, since *nāqidu* would not apply to bovines in the original usage." This distinction between the two terms is made already by A. Salonen (Review of W. von Soden, *Akkadisches Handwörterbuch*, *AOf* 23 [1970] 96): "*nāqidu* 'Hirt', d. h. das semitische Wort für ursprünglich 'Schafhirt' neben *rē'u* (*sic*) 'Rinderhirt.'" However, there is not a single attested Semitic language that distinguishes the terms in this way.

[52] M. Delcor, "Quelques termes relatifs à l'élevage des ovins en hébreu classique et dans les langues sémitiques voisines: étude de lexicographie comparée," in *Atti del*

dian dictionaries is very different from that of H. Waetzoldt in *Reallexikon der Assyriologie*: "sum. na-gada, akk. *nāqidu* . . . und gáb-ra, gáb-ús, akk. *kaparru* . . . ; beide hüten Groß- od. Kleinvieh. S. AHw und CAD."[53] In short, students of Hebrew have unjustly imposed their narrow, ovine interpretation of נוקד on Akkadian *nāqidu* and even attributed it to the Akkadian dictionaries. Only Tur-Sinai and Paul have defied this trend, using Akkadian to argue for a broader meaning of the Hebrew word.[54]

In view of the true meaning of the Akkadian cognate, it seems gratuitous to posit a contradiction between 1:1 and 7:14. It is more natural to assume that a בוקר is simply a specific type of נוקד. It is interesting to note that Driver, even though he held the narrowest possible view of the נוקד, resisted the temptation to posit such a contradiction: "From 7, 14 we learn that he had under his charge herds of larger cattle as well. . . ."[55] In light of the evidence presented above, the wisdom of Driver's approach should be evident.

As for the second question, involving the socioeconomic connotations of נוקד, modern scholars have generally returned to the view of Symmachus (according to one tradition),[56] the Targum,[57] and the early medieval exegetes[58] that the biblical נוקד was not a lowly hired hand.[59]

---

secondo congresso internazionale di linguistica camito-semitica (ed. P. Fronzaroli; Florence: Università di Firenze, 1978) 121-22.

[53] H. Waetzoldt, "Hirt. A. Philologisch (neusumerisch)," in *Reallexikon der Assyriologie* (ed. E. Ebeling, B. Meissner, et al.; Berlin/Leipzig: W. de Gruyter, 1928-) 4. 421-25. I am indebted to J. Huehnergard for this reference.

[54] N. H. Tur-Sinai, פשוטו של מקרא (Jerusalem: Kiryath Sefer, 1962-68) 3. 450; Paul, *Amos*, 35, 247-48, citing the Hammurabi passage. Danell does not deal explicitly with question (1), but his translations and discussions of the Akkadian and Hebrew terms in connection with questions (2) and (3) (*Amos*, 8-9) show that he too recognized the broader meaning. Cf. also D. Pardee, "The Baʿlu Myth," in *The Context of Scripture* (ed. W. W. Hallo and K. L. Younger; Leiden: E. J. Brill, 1997-2002) 1. 273 n. 283.

[55] Driver, *Introduction*, 293.

[56] See Origen, *Hexapla*, 1. 655 ἀρχιποιμήν "chief shepherd." So too, Ishodad of Merv (*Commentaire* [CSCO 303] 4. 82 line 7): רעיא ורש רעותא.

[57] See above.

[58] Rabbanites, e.g., Ibn Quraysh (רִסָאלָה, 276-77 no. 275) and Ibn Janāḥ (*ʾUṣūl*, 451 lines 5-7), and Karaites, e.g., Daniel al-Qumisi (פתרון, 32), Al-Fāsī (*Jāmiʿ al-Alfāẓ*, 2. 290 lines 21-22) and Yefet (Ms. British Library Or. 2400 = Margoliouth 282, p. קנו = f. 79b line 4). Cf. n. 3 above.

[59] By contrast, Jerome (*Commentarii*, 324 line 408) refers to Amos as a "humble and

This view is based first and foremost on the application of the term to Mesha, the king of Moab, in 2 Kgs 3:4.[60]

Many scholars have adduced Ugaritic evidence to support this view. Already in 1938 Montgomery wrote:

> The first published long text from Ras Shamra indicates that it is a "document" of a certain high official, who is *rb khnm*, "chief priest," *rb nqdm* "chief sheepmaster." This rare word appears to have had an official meaning, and Amos may have been more of "a gentleman" than critics have suspected.[61]

Subsequent discoveries of the term *nqd* in additional Ugaritic texts[62] and in two Akkadian texts from Ugarit have corroborated Montgomery's suggestion. A. F. Rainey writes: "The high status [of the *nqdm* at Ugarit] relative to other types of agricultural workers is indicated by the fact that they are the only class of agricultural workers that is included in the list of citizens holding feudal property."[63] P. C. Craigie develops this view further:

> [The *nqdm* in the Ugaritic texts] seem to have a higher status than mere labourers, or workers in general, as is indicated by the fact that under fiscal law a group of *nqdm* was equivalent to an entire village community. By way of contrast, the Ugaritic *rʿym* . . . , who are referred to in more than a dozen Ugaritic texts, appear to be closer to the level of labourers; they are described in various texts as receiving rations, work-

---

simple shepherd," and Abarbanel (*Comentario*, 20) calls him "the lowliest and poorest of shepherds."

[60] Interestingly, Moab was still known for its sheep in the Herodian period. According to *t. Men.* 9.13, the rams for the Temple service were imported from there; see S. Applebaum, "Economic Life in Palestine," in *The Jewish People in the First Century: Historical Geography, Political History, Social, Cultural and Religious Life and Institutions* (ed. S. Safrai and M. Stern; CRINT 1; Assen/Amsterdam: Van Gorcum, 1974-) 2. 647; S. Safrai, "The Temple," in *The Jewish People,* 2. 882; idem, "The Temple and the Divine Service," in *The World History of the Jewish People: The Herodian Period* (ed. M. Avi-Yonah; New Brunswick: Rutgers University Press, 1975) 321.

[61] J. A. Montgomery, "The New Sources of Knowledge," in *Record and Revelation* (ed. H. W. Robinson; Oxford: Clarendon Press, 1938) 22.

[62] See J.-L. Cunchillos and J.-P. Vita, *Concordancia de Palabras Ugaríticas* (Madrid-Zaragoza: Consejo Superior de Investigaciones Científicas, 1995) 1456; Olmo Lete and Sanmartín, *Diccionario* 2. 329-30 s.v. *nqd.*

[63] A. F. Rainey, מבנה החברה באוגרית (Jerusalem: Bialik, 1967) 86.

ing under supervision, and generally having a less elevated status than the *nqdm*. . . .[64]

. . . it is probably more accurate to think of Amos the *nōqēd* as being similar to the Ugaritic *nqd*; he probably owned, or managed, large herds of sheep and was engaged in the marketing of their products. Indeed, it was probably his marketing duties that took him north from his home state of Judah to the market towns of Israel, there to sell his goods. Taken together, the evidence indicates that Amos was engaged extensively in agricultural business, being involved in cattle and fruit-farming, in addition to sheep.[65]

Finally, Akkadian evidence has also been adduced. Many scholars, beginning with Engnell and Kapelrud,[66] have cited San Nicolò's demonstration that the *nāqidu*'s employed by the Eanna temple in sixth-century Uruk were "Viehzüchter" with a managerial role, the actual herding being done by *rē*ʾ*û*'s assigned to the *nāqidu*'s:

An der Spitze des mit der Aufzucht und Wartung von Rindern, Schafen und Ziegen des Tempels betrauten Personals standen ein oder mehrere ^{h}*rabi-bûli* "Viehvorsteher," in Eanna in der Regel drei, die meist selber Tierhalter waren. Jeder von ihnen hatte eine grössere Anzahl von "Viehzüchtern" (^{h}*nâqidu*, Ideogr. *NA.GAD*) unter sich. . . .

Soweit die Tiere draussen im Lande auf der Weide lebten, wurden sie von "Hirten" (^{h}*rê*ʾ*û*) unmittelbar betreut, die den einzelnen ^{h}*nâqidu* zugewiesen waren.[67]

Although biblical scholars continue to rely solely on San Nicolò's study from 1948, the subject has been revisited twice since then. Fortunately, San Nicolò's portrait of a three-tiered hierarchy at Uruk has been confirmed by these later investigations. H. M. Kümmel writes:

Soweit wir sehen können, entstammten die Viehvorsteher regelmäßig der Gruppe der ihnen unterstellten "Viehhalter," ^{lú}NA.GADA = *nāqidu*. . . .

Den Viehhaltern ihrerseits unterstanden die eigentlichen "Hirten" beim Vieh auf der Weide, ^{lú}SIPA = *rē*ʾ*û*. Das zeigen die gelegentlichen

[64] P. C. Craigie, "Amos the *nōqēd* in the Light of Ugaritic," *SR* 11 (1982) 32. See also M. Dietrich and O. Loretz, "Die ug. Berufsgruppe der *nqdm* und das Amt des *rb nqdm*," *UF* 9 (1977) 336-37.

[65] P. C. Craigie, *Ugarit and the Old Testament* (Grand Rapids: Eerdmans, 1983) 73.

[66] See I. Engnell, "Profetismens ursprung och uppkomst: Ett gammaltestamentligt grundproblem," *Religion och Bibel: Nathan Söderblom-Sällskapets Årsbok* 8 (1949) 15 n. 3; Kapelrud, *Central Ideas*, 6.

[67] San Nicolò, "Viehwirtschaft I," 284-85.

> Erwähnungen eines Hirten als PN_1, *rēʾû ša* PN_2 (*nāqidi*) "PN_1, Hirte des (Viehhalters) PN_2," bzw. PN_1 (*nāqidu*), PN_2 *rēʾû-šu* "der (Viehhalter) PN_1 (und) sein Hirte PN_2" deutlich.[68]

So too G. van Driel:

> The Uruk *nāqidu* is . . . sometimes a person with considerable holdings and connections. . . .
>
> These *nāqidu*'s were not themselves the actual herdsmen, the herding was left to the *rēʾû*'s.[69]
>
> If we assume the cattle herding was organised along lines similar to sheep herding, herdsmen contractually received a share of the young animals. . . .[70] Perhaps we could speculate on a 33% share for the herdsman, as is the case with sheep, though this would seem very generous with regards to cattle.
>
> . . . The *nāqidu*'s, the administrative herdsmen, will have been managers. We can only assume that part of their herds was their personal property.[71]

The status of the Old Babylonian *nāqidu(m)* and the Sumerian na-gada is far less clear. There is no evidence that the *nāqidu(m)*/na-gada was superior to the *rēʾû(m)*/sipa. Nonetheless, some scholars believe that the Sumerian term na-gada referred, at least sometimes, to an *Oberhirt*.[72] The subordinate of the *nāqidu(m)*/na-gada, the one who did the actual herding, was the *kaparru(m)*/gáb-ra.[73]

[68] Kümmel, *Familie*, 49-50. According to San Nicolò ("Viehwirtschaft I," 285), there was also an "Oberhirt" (*rab-rēʾî*) who had a supervisory role; Kümmel (*Familie*, 50) notes that, at Uruk, there is only a single, very uncertain attestation of this term.

[69] G. van Driel, "Neo-Babylonian Sheep and Goats," *Bulletin on Sumerian Agriculture* 7 (1993) 224-25.

[70] This is also what Jacob received from Laban, according to Genesis 30-31; see further J. J. Finkelstein, "An Old Babylonian Herding Contract and Genesis 31:38f," *JAOS* 88 (1968) 30-36.

[71] Driel, "Cattle," 228-29.

[72] See Landsberger, *Materialien*, 106 and F. R. Kraus, *Staatliche Viehhaltung im altbabylonischen Lande Larsa* (Amsterdam: Noord-Hollandsche U. M., 1966) 16, 50; the latter compares the Roman *magister pecoris*, who had to be able to write and keep books. J. N. Postgate ("Some Old Babylonian Shepherds and Their Flocks," *JSS* 20 [1975] 2) appears to agree. By contrast, Van De Mieroop ("Sheep and Goat," 168) states that the term na-gada refers to the lowest level in the hierarchy of herdsmen.

[73] See Finkelstein, "Herding Contract," 32 n. 6; M. W. Green, "Animal Husbandry at Uruk in the Archaic Period," *JNES* 39 (1980) 16 n. 82; Waetzoldt, "Hirt," 421. The first

Socioeconomic conclusions can be drawn not only from the term *nāqidu*/נוקד, but also from Amos' association with both צאן and בקר, if we may judge once again from the situation in Uruk:

> . . . there are no flocks of sheep and goats in combination with cattle. When people are held accountable for both categories they belong to what might be called managerial levels.[74]

We come finally to the claim that the term נוקד had sacral connotations. This claim appears to have arisen in response to the discovery and decipherment of the tablets from Ras Shamra. In the colophon of the Baal cycle (UT 62:56, CTA 6 VI:56, KTU 1.6 VI:56), discovered in 1933 and published in 1934 by C. Virolleaud, a single individual is identified as *rb khnm rb nqdm* "chief of the priests and chief of the herdsmen."[75] Virolleaud found this association surprising at first glance, but concluded that it was only natural for sheep-breeding to be a prestigious occupation in an agricultural society such as that at Ugarit.[76] In the following year, T. H. Gaster suggested reading more into the association:

> Virolleaud opines that *rb nḳdm* refers to the fact that the high-priest was a sheep-breeder, as were many people at Ugarit.
>
> May not the title have a special meaning and refer to the sacred sheep of the temple?[77]

In 1943, I. Engnell cited the Ugaritic phrase as evidence that the king in the ancient Near East was the high priest *par excellence*:

> For another district, too, *viz.* Moab, the O. T. has something to say, 2 Ki. 3.4 styling king Meša as נקד. One has for a long time—owing to the

two scholars seem to treat *rēʾû(m)* as an approximate synonym of *nāqidu(m)*; the last believes that *rēʾû(m)* is a superordinate term, covering both the *nāqidu(m)* and the *kaparru(m)*.

[74] Driel, "Sheep and Goats," 219.

[75] For a recent discussion of the colophon, see Pardee, "Baʿlu," 273 nn. 281-83.

[76] C. Virolleaud, "Fragment nouveau du poème de Môt et Aleyn-Baal," *Syria* 15 (1934) 242-43.

[77] T. H. Gaster, "Notes on Ras Shamra Texts," *OLZ* 38 (1935) 475. For a different explanation of the association from around the same time, see E. Dhorme, *L'évolution religieuse d'Israël* (Bruxelles: Nouvelle société d'éditions, 1937) 223.

> Arabic and Accadian etymologies—suspected this "shepherdship" to be of a sacral nature; and now the Ras Shamra texts have supplied the final proof. Gaster seems to be the first to have observed that נקד occurs at R Sh as a sacral term in the colophon of I AB, where רב נקדם is paralleled with רב כהנם and, evidently, also with ראש רעי. Hence *nōḳēd* must denote king Meša as a sacral person, and probably as the high priest in principle.[78]

As for the other biblical נוקד, Engnell wrote:

> The import of the word נקד, now established through the R Sh parallel, ought to have its consequences also for the conception of Amos' person and position (Am. 1.1). The present writer hopes to have the opportunity of reverting to the Amos question.[79]

Engnell's reference to Arabic and Akkadian etymologies is perplexing and his reference to Gaster, misleading. Gaster's modest suggestion was that the *chief* of the *nqdm* at Ugarit raised sacred *sheep*. It is a long way from that to Engnell's theory that *every nqd everywhere* in the ancient Near East was a sacral *person*. A less dramatic leap was made in 1948 by D. M. L. Urie: "At Ugarit the *nqd* was obviously a cult official."[80]

The consequences for Amos of Engnell's theory were discussed first by his Uppsala colleague, A. Haldar, in 1945. Haldar pointed to the term נוקד as evidence that Amos and indeed all of the writing prophets were cultic functionaries, members of cultic associations headed by the king.[81] Engnell himself returned to the question in 1948:

> [Amos] is reported . . . to have belonged to "the herdsmen from Tekoa," a place a couple of miles from Jerusalem. The Hebrew word for "herdsmen" in this case, *nōḳēd*, reveals, upon comparison with corresponding

[78] I. Engnell, *Studies in Divine Kingship in the Ancient Near East* (Uppsala: Almqvist & Wiksell, 1943) 87.

[79] Engnell, *Studies*, 87 n. 2.

[80] D. M. L. Urie, "Officials of the Cult at Ugarit," *PEQ* 80 (1948) 46. His view is cited approvingly in A. DeGuglielmo, "Sacrifice in the Ugaritic Texts," *CBQ* 17 (1955) 89 and J.-M. de Tarragon, *Le Culte à Ugarit* (Paris: J. Gabalda, 1980) 135.

[81] A. Haldar, *Associations of Cult Prophets Among the Ancient Semites* (Uppsala: Almqvist & Wiksell, 1945) 112 n. 4.

> terms in other Semitic languages, something that the Ras Shamra texts also show, that A. had, in some way, a connection with a personnel class tied to the temple. Tekoa must have been a "branch" of the temple in Jerusalem. In view of the way in which, in antiquity, all economic and political life was centralized in the sanctuaries, his position as "herdsman" and, at the same time, member of a temple personnel group of some kind can be easily understood.[82]

Not all of Engnell's Swedish colleagues embraced his view of נוקד as a sacral term. In an article published in 1949, E. Sjöberg attacked Engnell's reasoning, arguing that the existence of one cultic *nqd* at Ugarit did not prove that all *nqdm* were cultic herdsmen, a point made also by E. Würthwein in responding to Haldar in 1950.[83] Sjöberg claimed that, in Mesopotamia, "just as there were cultic *nā-ḳi-du*, there were ordinary, profane ones, who could be placed together with cowherds . . . and farmers . . . into one group."[84] Furthermore, "the Arabic word *naqad*, from which the Hebrew word is, for the most part, customarily derived has no cultic relation whatsoever. It is quite simply a designation of a certain species of sheep."[85] Finally, "the juxtaposition [of herding and growing sycomore figs] becomes more natural, if in both cases it is a question of an ordinary middle-class food—if the sheep were as non-cultic as the sycomore figs."[86]

Engnell's sole response to this refutation of his thesis was to cite the article by San Nicolò that was to become a cornerstone of the debate.[87] This response was essentially beside the point, since Sjöberg had already conceded that "there were cultic *nā-ḳi-du*" in Mesopotamia.

In 1951, there were further demurrals in Sweden and Germany, from G. A. Danell and O. Eissfeldt.[88] In that same year, however, M. Bič

[82] I. Engnell, "Amos," in *Svenskt Bibliskt Uppslagsverk* (Gävle: Skolförlaget, 1948) 1. 59-60.

[83] E. Sjöberg, "De förexiliska profeternas förkunnelse," *SEÅ* 14 (1949) 16; E. Würthwein, "Amos-Studien," *ZAW* 62 (1950) 39 n. 59.

[84] Sjöberg, "Förkunnelse," 17 n. 14.

[85] Sjöberg, "Förkunnelse," 17 n. 14.

[86] Sjöberg, "Förkunnelse," 18.

[87] San Nicolò, "Viehwirtschaft I," is cited in Engnell, "Profetismens ursprung," 15 n. 3.

[88] G. A. Danell, "Var Amos verkligen en nabi?" *SEÅ* 16 (1951) 8-9; O. Eissfeldt, "The Prophetic Literature," in *The Old Testament and Modern Study* (ed. H. H. Rowley; Oxford: Oxford University Press, 1951) 159, cf. 124.

took Engnell's theory a step further, claiming that the real meaning of the allegedly sacral term *nāqidu*/נוקד is "hepatoscopist."[89] This curious claim was immediately refuted by Murtonen.[90] When the theory reappeared in the following decade in the work of J. Gray, it was refuted again, by Segert.[91] Bič argued once again for his claim in 1969, this time eliciting a refutation by Wright.[92]

Engnell's theory was popularized by A. S. Kapelrud.[93] The latter spelled out evidence that the former only hinted at, and he formulated the theory in a more cautious manner:

> Who, then, was a *noqed* in Judah in the time of Amos? It may have been a person of rather high rank who was responsible for a large part of the temple herds. Economically, as well as in what concerned the temple cult, he was therefore an important person. . . .[94]

> He may . . . officially have had something to do with the cult, even if his task has only been to furnish it with the necessary sheep for sacrifices.[95]

Kapelrud's other contribution was to adduce additional Ugaritic evidence:

> In text No. 113 (Gordon) *nqdm* are listed together with yeomen, *ṯnnm*, priests, *khnm*, and another class of priests, *qdšm* (vv. 70—73). In No. 300 we find *nqdm* listed after *ṯġrm*, door-men, gatekeepers, and *šrm*, singers.

[89] M. Bič, "Der Prophet Amos—Ein Haepatoskopos," *VT* 1 (1951) 293-96.

[90] A. Murtonen, "The Prophet Amos—A Hepatoscoper?" *VT* 2 (1952) 170-71.

[91] S. Segert, "Zur Bedeutung des Wortes nōqēd." *Hebräische Wortforschung: Festschrift zum 80. Geburtstag von Walter Baumgartner* (VTSup 16; Leiden: E. J. Brill, 1967) 279-83.

[92] M. Bič, *Das Buch Amos* (Berlin: Evangelische Verlagsanstalt, 1969) 8-9, 16-20; Wright, "Liver," 3-6.

[93] Kapelrud, *Central Ideas*, 6. Kapelrud is frequently assumed to be the originator of the theory, because he failed to attribute it to Engnell. The Swedish encyclopedia entry in which Engnell published the theory is not well known to most scholars. Kapelrud cites it frequently in his book but not in this context. The only work that Kapelrud cites in arguing for a connection of the *nqdm* with the temple is Montgomery, "The New Sources," quoted above, which says nothing about such a connection. Even stranger, Engnell himself credits Montgomery, "Notes," but that too does not deal with question (3) in any way, as noted already by Sjöberg, "Förkunnelse," 17-18 n. 15.

[94] Kapelrud, *Central Ideas*, 6.

[95] Kapelrud, *Central Ideas*, 69. See further below.

> . . . There can be little if any doubt at all that the *nqdm* are mentioned among the temple personnel because they were an important guild in the service of the temple.[96]

UT 113:71 (CTA 71:71, KTU 4.68:71) had already been discussed by Sjöberg in 1949, when the meaning of the word *ṯnnm* was still considered uncertain. Since the *ṯnnm* are grouped in that text with the *nqdm*, Sjöberg felt that the precise nuance (cultic or non-cultic) of the word *nqdm* was also uncertain.[97] For H. J. Stoebe, writing eight years later, the situation seemed far more clear-cut. In his view, Kapelrud's interpretation of Text 113 was very dubious:

> Die nqdm sind mit den ṯnnm zusammengefaßt und haben mit diesen zusammen einen Bogenschützen zu stellen; ṯnnm stellen vielleicht ebenso wie die vorher genannten mrum militärische Klassen kar. Es muß also sehr zweifelhaft sein, ob die nqdm hier mit den khnm und den qdšm zusammenzustellen sind. . . .[98]

Stoebe also impeached the evidence of UT 300 rev:12 (CTA 82 B:12, KTU 4.103:44):

> Indessen werden gerade in diesem Zusammenhang khnm and qdšm nicht aufgeführt, außerdem werden zu Anfang des Verzeichnisses Berufsgruppen mit Land dotiert, die kaum zum Tempelpersonal zu rechnen sind, so daß das Prinzip der Anordnung dieser Liste einigermaßen dunkel bleibt.[99]

Finally, Stoebe questioned whether the *nāqidu*'s employed by the Eanna temple are to be considered true cult personnel:

> . . . aber es ist damit wohl noch nicht entschieden, daß diese Leute den eigentlichen Kultpersonen des Tempels zuzurechnen sind. Trotz einer unbestreitbaren Abhängigkeit vom Tempel können sie doch ein gewisses Maß an Selbständigkeit gehabt haben, ja müssen es wohl sogar gehabt

[96] Kapelrud, *Central Ideas*, 6.
[97] Sjöberg, "Förkunnelse," 16-17 n. 14.
[98] Stoebe, "Der Prophet Amos," 166.
[99] Stoebe, "Der Prophet Amos," 167.

haben, weil die Weideplätze oft in erheblicher Entfernung von den Tempeln lagen.[100]

Stoebe's distinction between temple personnel and cult personnel is standard in recent work. Wolff writes:

> To be sure, it is quite possible that at Ugarit, as at Babylon, sheep breeders were responsible for temple flocks. But must they therefore have belonged to the cultic personnel? Certainly in the case of Amos such a positive conclusion should not be drawn.[101]

So too Craigie:

> Thus, the term *nqd* does not necessarily carry any sacral or religious connotations. While the *nqd* could be a temple-servant, the majority of the evidence indicates that the *nqdm* were servants of the royal establishment. And even in the single instance in which temple-*nqdm* can be identified, the text gives no indication whatever that their role was in any sense sacral.
>
> Applying this information to the role of Amos, the *nqd*, it is clear that while Amos could have been a temple-servant, it is far more likely that he was not.[102]

A similar conclusion is reached by B. Cutler and J. Macdonald in their discussion of the *nqdm* of UT 113:

> [Sheep-breeders] would have been needed to meet the unending needs of the palace with its wide repertoire of victuals for multivarious guests from different countries, as well as the king's personal hospitality to men of rank. . . . Thus Meshaᶜ and Amos would have been prosperous business men before (and during?) their respective roles of rule and prophecy.[103]

[100] Stoebe, "Der Prophet Amos," 166; cf. also 174-75.

[101] H. W. Wolff, *Joel and Amos* (trans. W. Janzen et al.; ed. S. D. McBride, Jr.; Hermeneia; Philadelphia: Fortress, 1977) 124.

[102] Craigie, "Amos," 33.

[103] B. Cutler and J. Macdonald, "The Unique Ugaritic Text UT 113 and the Question of 'Guilds,'" *UF* 9 (1977) 25. Cf. also B. Cutler and J. Macdonald, "Identification of the *Naᶜar* in the Ugaritic Texts," *UF* 8 (1976) 31 n. 30; and M. Heltzer, "Royal Economy in Ancient Ugarit," in *State and Temple Economy in the Ancient Near East, II* (ed. E. Lipiński; Orientalia Lovaniensia Analecta 6; Leuven: Departement Oriëntalistiek, 1979) 472: "Among the people engaged in royal cattle-breeding, one finds, first of all, different kinds of shepherds, designated by the terms *nqdm* and *rᶜym*."

As for the *ṯnnm*, who are grouped in that text and several others with the *nqdm*,[104] Cutler and Macdonald note that the equivalent group at Alalakh, the *šanannu*, appear as owners of livestock in several Akkadian texts. They conclude:

> There can be little doubt, therefore, that the *šanannu* were stockholders who pastured their sheep in grass (Texts 341, 350), . . . and who played their role as military men only when required to do so in accordance with treaty agreements with the overlord of the territory in which they grazed their animals and, presumably, to whom they sold their sheep.[105]

The most recent work on the Eanna herdsmen has made the notion of the cultic *nāqidu* even less plausible. Van Driel stresses the importance of a distinction that was not made explicit by either San Nicolò or Kümmel:

> We must differentiate between that part of the documentation belonging to the flocks directly managed by the temple personnel and its administrative supervisors, and the flocks managed indirectly, through written, and possibly unwritten, contract.[106]

> There is an internal organization which provides the animals required for the cult functions along different lines from the external organization, which, at least in part operates at considerable distances from the towns where the institutions have their abode. Especially in Uruk it is obvious that (some? of) the external herding was contracted out. . . .
>
> The fundamental difference is that the personnel of the "home herds" figures in the ration lists, whereas the extramural personnel does not.[107]

According to van Driel, the Uruk *nāqidu* was part of the *external* organization:

> In Uruk the *nāqidu* was a person with sometimes wide ranging interests in cattle and sheep herding and in arable farming. As an entrepreneur he had aquired [*sic*] a position between the temple-administration and its herds in the external organization of sheep breeding.[108]

104 Cunchillos and Vita, *Concordancia*, 1456.
105 Cutler and Macdonald, "UT 113," 26.
106 Driel, "Sheep and Goats," 219.
107 Driel, "Sheep and Goats," 224.
108 Driel, "Sheep and Goats," 225.

It is, therefore, arguable that the *nāqidu*'s hired by the Eanna temple were not even temple personnel, let alone cult personnel. If any of the Eanna herdsmen fit Engnell's description, it is the herdsmen of the *internal* organization, but they have a different title: *rēʾû ginê* or *rēʾû sattukki*.[109] In short, there is no longer the slightest basis for Engnell's claim that the term נוקד had sacral overtones.

It remains to be said that the phrase actually used in Amos 1:1 is not היה נקד (as in 2 Kgs 3:4) but היה בנקדים. It would appear that a term like בנקדים implies membership in an exclusive group. That is the case in the following parallels: כל זכר בכהנים יאכל אתה "any male among the priests may eat it" (Lev 6:22); הגם שאול בנביאים "Is Saul too among the prophets?" (1 Sam 10:11); אחיתפל בקשרים עם אבשלום "Ahithophel is among the conspirators with Absalom" (2 Sam 15:31); והיו באכלי שלחנך "and let them (the sons of Barzillai the Gileadite) be among those that eat at your table" (1 Kgs 2:7); and והמה בגבורים עזרי המלחמה "they were among the warriors who gave support in battle" (1 Chr 12:1). We shall investigate the nature of the group to which Amos belonged in the next section and in chapter 5 below.

The Syntax of דברי עמוס אשר היה בנקדים מתקוע

According to the Peshiṭta, Vulgate, most medieval exegetes, and many modern scholars, מתקוע modifies נקדים rather than עמוס.[110] If so, Amos 1:1 contains a reference to "the herdsmen from Tekoa." This syntactic analysis was challenged in the nineteenth century by H. Oort and K. Budde, based on the evidence of Jer 1:1: דברי ירמיהו בן חלקיהו מן הכהנים אשר בענתות "the words of Jeremiah son of Hilkiah, of the priests that were in Anathoth." In the words of Budde:

> Der vorliegende Wortlaut אשר היה בנקדים מתקוע kann überhaupt nicht lediglich dazu dienen sollen, des Amos Heimat und Stand anzugeben. Das würde, wie Oort richtig hervorhebt, nach Jer. 1, 1 lauten müssen מן הנקדים אשר בתקוע. . . . Was hier steht, könnte etwa heissen, dass Amos zu einer Schaar von Viehzüchtern aus Tekhoa gehört habe, die sich zu irgend einer bestimmten Zeit an einem anderen Orte einfanden oder aufhielten, so etwa, wie sich bei der Belagerung Jerusalem's die Reka-

[109] Driel, "Sheep and Goats," 226.
[110] Weiss, ספר עמוס, 2. 8 n. 77.

> biten hinter die Mauern der Hauptstadt flüchteten (Jer. 35). Da aber eine solche Gelegenheit nicht zu ersinnen, noch weniger genannt ist, kann diese Auffassung nicht in Betracht kommen.[111]

According to Budde, the parallel in Jer 1:1 would lead one to expect "the herdsmen *in* Tekoa" instead of "the herdsmen *from* Tekoa." The latter formulation would make sense if the herdsmen were outside of Tekoa (e.g., taking refuge in Jerusalem), but, since that was not the case in Budde's view, the interpretation is impossible. Budde's solution is to take "from Tekoa" as modifying "Amos":

> . . . so erklärt sich das schwierige מן von selbst. Es ist eben das מן der Herkunft, der Heimat, unmittelbar an den Eigennamen anschliessend, wie אבצן מבית לחם Richt. 12, 8, vgl. Kön. II, 21, 19. 23, 36, gleichbedeutend mit dem Gentilicium התקועי, das sich in Ueberschriften von Propheten-büchern in מיכה המרשתי und נחום האלקשי findet.[112]

Budde's argument has been widely accepted, especially in Germany;[113] however, it suffers from a number of weaknesses. First, Budde seems to have assumed that his "מן of origin" is used only with personal names. But, in fact, there are examples with common nouns, e.g., איש מהר אפרים (Judg 17:1), נער מבית לחם יהודה (Judg 17:7), איש מבנימין (1 Sam 17:27), אנשים מאשר ומנשה ומזבלון (2 Chr 30:11). Second, Budde ignored the explanation for the phrase "from Tekoa" that had already been offered by F. Hitzig. According to Hitzig, the work of the Tekoite herdsmen took them out of town, to the grazing lands of מדבר תקוע (2 Chr 20:20), where they stayed with their animals.[114] At Nuzi too:

> The pasturing of the flocks and herds took the herdsmen away from settled areas. Moreover, other activities such as the counting, shearing, plucking and slaughter of livestock probably occurred in agricultural

[111] K. Budde, "Die Ueberschrift des Buches Amos und des Propheten Heimat," in *Semitic Studies in Memory of Rev. Dr. Alexander Kohut* (ed. G. A. Kohut; Berlin: S. Calvary, 1897) 107.

[112] Budde, "Die Ueberschrift," 109.

[113] See H. F. Fuhs, "Amos 1,1: Erwägungen zur Tradition und Redaktion des Amosbuches," in *Bausteine biblischer Theologie: Festgabe für G. Johannes Botterweck* (ed. H.-J. Fabry; Köln-Bonn: P. Hanstein, 1977) 275 n. 16.

[114] F. Hitzig, *Kleinen Propheten* (4th ed.; Leipzig: S. Hirzel, 1881) 108.

areas surrounding the cities. One has only to remember the shepherd's hut in Gilgamesh to envision the herdsmen's normal environment.[115]

In all likelihood, the Tekoite herdsmen spent part of the year even further from home (e.g., in the Jericho Valley); seasonal migration (transhumance) would have been unavoidable, since מדבר תקוע does not have pasturage throughout the year.[116] All in all, the Tekoite herdsmen probably spent very little time at home with their families, and thus "*from* Tekoa" is a far more natural description of them than "*in* Tekoa."

It should also be noted that the superscription may be using a Jerusalemite expression or, at least, expressing a Jerusalemite perspective.[117] The Tekoites would have been familiar figures in the Jerusalem livestock market, which, in view of its proximity and size, was presumably their main outlet.[118] It is quite possible that in Jerusalem they were known popularly as "the herdsmen from Tekoa" to distinguish them from, say, "the herdsmen from Hebron" or "the herdsmen from Moab." For shoppers looking for their stall in the Jerusalem market, it would make little sense to inquire about the whereabouts of the "herdsmen *in* Tekoa."

Finally, the assumption that מתקוע modifies עמוס creates a very strange sequence of attributive modifiers: an asyndetic prepositional phrase sandwiched between two syndetic relative clauses. The normal order for these modifiers would be: מתקוע, **אשר** היה בנקדים, **אשר** חזה על ישראל בימי עזיה מלך יהודה ובימי ירבעם בן יואש מלך ישראל שנתים לפני הרעש. To account for this anomaly, Budde was forced to make the further assumption that the phrase אשר היה בנקדים is a later insertion, perhaps from the margin. Budde attempted to provide additional motivation for this assumption, unsuccessfully in my opinion.[119]

[115] M. A. Morrison, "Evidence for Herdsmen and Animal Husbandry in the Nuzi Documents," in *Studies on the Civilization and Culture of Nuzi and the Hurrians* (ed. M. A. Morrison and D. I. Owen; Winona Lake, Ind.: Eisenbrauns, 1981-) 1. 259 n. 14.

[116] See further in chapter 5 below.

[117] For the origin of this superscription and its connections to Jerusalem, see D. N. Freedman, "Headings in the Books of the Eighth-century Prophets," *AUSS* 25 (1987) 9-26.

[118] See further in chapter 5 below.

[119] Budde's argument ("Die Ueberschrift," 110) is well summarized by Wolff (*Joel*

In my view, the traditional syntactic analysis of Amos 1:1 is preferable, for the reasons mentioned above and for one additional reason. There seems to be another reference in the Bible to the herdsmen of Tekoa that has been almost completely ignored in discussions of our verse.[120] As recognized by A. B. Ehrlich, the נקדים of Tekoa are probably the אדירים of Tekoa mentioned in Neh 3:5 ועל ידם החזיקו התקועים ואדיריהם לא הביאו צַוָּרם בעבדת אדניהם "and next to them, the Tekoites repaired, but their אדירים did not take upon their shoulders the work of their lord." According to Ehrlich, אדירים in this verse refers to wealthy shepherds and is equivalent to the fuller expression אדירי הצאן in Jer 25:34-36 (which, in turn, stands in opposition to צעירי הצאן in Jer 49:20).[121]

A similar interpretation of אדירי הצאן is found already in Ibn Janāḥ's dictionary: "great shepherds."[122] According to Ibn Janāḥ, אדירים has the same meaning in Judg 5:25. Tur-Sinai's view is similar, except that he sees no implication of wealth or greatness: "אדיר is equivalent to רועה, נוקד."[123] He adduces additional evidence for this interpretation, e.g., the parallelism between רעיך and אדיריך in Nah 3:18.[124]

---

and Amos, 117): "While the first אשר-clause unquestionably has 'Amos' as its antecedent, the second probably refers back to the 'words of Amos' (דברי עמוס), since the comparable superscriptions (Isa 1:1; 2:1; 13:1; Mic 1:1; Hab 1:1) always specify an object for the verb 'to view' (חזה). Two relative clauses with such different antecedents hardly flowed from the same pen." In fact, the difference in the antecedents posited by Budde and Wolff is by no means unusual. The construction that Wolff finds in our verse is just a special case of a well-attested construction in Biblical Hebrew (and Arabic): an attributive modifier of a genitive noun followed by an attributive modifier of the entire genitive phrase; see R. C. Steiner, "Ancient Hebrew," in *The Semitic Languages* (ed. R. Hetzron; London: Routledge, 1997) 165. Two examples with genitive phrases of the form דברי X are Deut 5:24 קול דברי העם הזה אשר דברו אליך "the sound of the words of this people which they spoke to you" and 28:58 את כל דברי התורה הזאת הכתובים בספר הזה "all the words of this Teaching that are written in this book."

[120] There is a fleeting allusion to this reference in Y. Ziv, "בוקר ובולס־שקמים"—בתקוע?, *Bet Mikra* 92 (1982-83) 50 n. 16.

[121] A. B. Ehrlich, מקרא כפשוטו (Berlin: M. Poppelauer, 1899-1901) 2. 418.

[122] Ibn Janāḥ, *ʾUṣūl*, 22 line 17.

[123] N. H. Tur-Sinai, בשולי המלון של אליעזר בן־יהודה, *Leš* 13 (1944-45) 108; idem, הלשון והספר: כרך הלשון (2nd ed.; Jerusalem: Bialik, 1954) 444. So too Saadia Ibn Danān, *Sefer ha-šorašim* (ed. M. J. Sánchez; Granada: Universidad de Granada, 1996) 32 s.v.

[124] Tur-Sinai, בשולי המילון, 108; idem, הלשון והספר: כרך הלשון, 444.

The Meaning of ויקחני ה׳ מאחרי הצאן

If we are right in assuming that Amos was a *nāqidu ša ṣēni u lâti*, dealing with both צאן and בקר, then the contradiction between 7:15 (מאחרי הצאן) and 7:14 (בוקר) is only apparent. But even if there is no logical problem, there is a problem on the level of conversational implicature: the singling out of בקר in 7:14 and צאן in 7:15 still requires explanation. We have already noted that singling out בקר makes perfect sense in 7:14, where the emphasis is on Amos' financial self-sufficiency. By the same token, singling out צאן makes perfect sense in 7:15, where the emphasis is on Amos' legitimacy as a leader. Andersen and Freedman write:

> [Amos'] mandate came from God himself, who took him from following the flock—a cliché out of Israel's past but one that was packed with tradition and power. Israel's history was largely shaped by ex-shepherds: Moses, who was caring for a flock when summoned directly to service by the God of the holy mountain; and David, the archetypal shepherd boy, who was called to be the Lord's anointed from his duties to the flock to serve a larger flock as ruler and king.[125]

A similar idea was articulated already in the fifteenth century by Abarbanel:

> And the shepherds of Israel and its leaders were all herders of צאן, as is apparent from the patriarchs and Moses and David, on account of that trade being similar to leading the people. That is why he (Amos) says ויקחני ה׳ מאחרי הצאן and does not mention the בקר. . . .[126]

This analysis accounts for the singling out of צאן in 7:15 by comparing Amos with a restricted set of leaders: Moses and David and perhaps the patriarchs. But what about the leaders who were not ex-shepherds, e.g., Saul, Elisha, and Gideon? One might reasonably argue that those other leaders are counterexamples and that their exclusion from consideration is arbitrary.[127]

[125] Andersen and Freedman, *Amos*, 790.

[126] Abarbanel, *Comentario*, 196.

[127] For an extreme example of this broader approach, see the list of nineteen parallels (including a 20th-century German politician) in Schult, "Legitimation," 463-74.

The only objective criterion for excluding those leaders compels us to exclude Moses and the patriarchs as well. I am referring to the striking similarity between ויקחני ה׳ מאחרי הצאן . . . אל עמי ישראל in Amos 7:15 and אני לקחתיך מן הנוה מאחר הצאן . . . על עמי על ישראל in 2 Sam 7:8. Neither לקח ה׳ את X מאחר(י) הצאן nor לקח ה׳ את X מאחר(י) Y, appears to be a stereotyped formula. They are never used of any other biblical figure, even though they seem appropriate to some of them. Moses never says ויקחני ה׳ מאחרי הצאן, even though that description would seem to fit him at least as well as Amos. His call in Exod 3:10 comes when he is literally behind his flock; cf. Exod 3:1: וינהג את הצאן "he drove the צאן (from behind)." Elisha never says ויקחני ה׳ מאחרי הבקר, even though he was literally behind a team of oxen at the moment of his call in 1 Kgs 19:19. It should not be assumed that אחרי was the only preposition that could have been used in these expressions. 1 Sam 16:19 employs a different preposition: שלחה אלי את דוד בנך אשר בצאן "send me your son David, who is with (lit. among) the צאן."

H. Schult recognizes that the similarity between Amos 7:15 and 2 Sam 7:8 cannot be due to chance, and he admits that an explanation positing direct dependence of Amos 7:15 on 2 Sam 7:8 (quotation, allusion or the like) would, in theory, be more solid than one positing a looser connection. In practice, however, such an explanation is impossible "weil es einleuchtende Gründe für eine theologische oder sonst überlieferungsgeschichtliche Verbindung von Amos mit David gar nicht gibt." Even "die Nähe von Am 9_{11} ('zerfallene Hütte Davids') zur Nathanweissagung" does not solve this problem, according to Schult. It does not help us to understand "was David und Amos als Personen oder Gestalten nach Auffassung der Tradition miteinander verbinden soll."[128] M. Weiss accepts this conclusion:

> It is impossible to imagine what could have been the motivation, conscious or unconscious, for formulating the words of Amos concerning his being taken to prophesy in the language of the word of the Lord concerning David being taken to rule.[129]

Three objections may be raised against the discussions of Schult and Weiss. First, they do not mention that the expression לקח ה׳ את X

[128] Schult, "Legitimation," 476-78, esp. 477.

[129] Weiss, ספר עמוס, 1. 240.

מאחר(י) הצאן is used of David not only in 2 Sam 7:8 but also in Ps 78:70-71, where it is distributed over two hemistichs: ויבחר בדוד עבדו *ויקחהו* ממכלאת צאן: *מאחר* עלות הביאו.[130] Also worthy of mention is 11QPs[a] 151, in which David says: וישלח ויקחני מאחר הצואן.[131] Clearly this phrase was a stock expression associated with David.

Second, Schult's passing reference to "die Nähe von Am 9_{11} ('zerfallene Hütte Davids') zur Nathanweissagung" scarcely does justice to the topic. Here too echoes of 2 Sam 7 have been noted.[132] The divine promise concerning "David's booth" in Amos 9:11 (ביום ההוא *אקים* את סכת דויד הנפלת . . . *ובניתיה כימי עולם*) echoes the ideas and, to some extent, the language of the divine promise concerning David's "house" in Nathan's oracle (as recorded in 2 Sam 7:11-16 and as cited in 2 Sam 7:27): בית יעשה לך ה׳ (2 Sam 7:11), *והקימתי* את זרעת אחריך (7:12), ונאמן ביתך וממלכתך עד *עולם* (7:16), בית *אבנה* לך (7:27).[133] Each of these echoes functions as what B. D. Sommer, following Z. Ben-Porat, calls "a *marker*, an identifiable element or pattern in one text belonging to another independent text."[134] According to Sommer, an "abundance

[130] See E. Z. Melamed, "The Breakup of Stereotyped Phrases as an Artistic Device in Biblical Poetry," *Scripta Hierosolymitana* 8 (1961) 115-53.

[131] J. A. Sanders, *The Psalms Scroll of Qumrân Cave 11* (DJD IV; Oxford: Clarendon Press, 1965) 55. I am indebted to M. S. Smith for reminding me of this parallel.

[132] See Wolff, *Joel and Amos*, 353.

[133] "David's booth" is apparently the dilapidated remnant of David's "house." Precisely what it refers to has been the subject of much debate. One piece of evidence seems to have been ignored. In Amos 9:12, סכת דויד takes a plural verb (assuming that the subject of יירשו is not אשר נקרא שמי עליהם), just as גוי takes a plural verb in Amos 6:14 (כי הנני מקים עליכם . . . גוי ולחצו אתכם) and בית דוד takes a plural verb in Isa 7:13 (שמעו נא בית דוד). This suggests that סכת דויד is a collective, referring to a group of people. For the switch from singular (בניתיה) to plural (יירשו), cf. the treatment of גוי/עם in Jer 6:22-24 and the examples cited in Steiner, "Ancient Hebrew," 166.

[134] B. D. Sommer, *A Prophet Reads Scripture: Allusion in Isaiah 40-66* (Contraversions; Stanford: Stanford University Press, 1998) 11. Recognition of the marker(s) is the first of 3-4 steps in "actualizing" an allusion. As noted by R. Alter ("Putting Together Biblical Narrative," Bilgray lecture, University of Arizona, 10 March 1988, 2), biblical scholarship has traditionally, but unjustifiably, neglected allusion:

> Allusion to antecedent literary texts is an indispensable mechanism of all literature, virtually dictated by the self-recapitulative logic of literary expression. No one writes a poem or a story without some awareness of other poems or stories to emulate, pay homage to, vie with, criticize, or parody, and so the evocation of phrases, images, motifs, situations from antecedent texts is an essential part of the business of making new texts. For reasons that I hope will soon be clearer, the

of markers pointing back to the older text makes clear that [the author] borrowed from that text," unless "both [texts] utilize stock vocabulary, exemplify a literary form such as a lament, or treat a subject that calls for certain words."[135] Based on this criterion, it is reasonable to conclude that Amos made use of Nathan's oracle.[136]

Third, Schult and Weiss are too hasty in concluding that it is impossible to explain why Amos would be alluding to David in 7:15. It is possible that Amos uses the phrase ויקחני ה' מאחרי הצאן in order to associate himself with David, in opposition to Amaziah and Jeroboam. Like David, he is a legitimate leader, taken away from following the flock by God.[137] Neither Amaziah nor his king, Jeroboam, can make that claim.

In any event, it is now apparent that the reference to צאן in 7:15 was necessitated by the fixed form of the expression לקח ה' את X מאחר(י) הצאן. This is true whether we are dealing with an allusion or a formula. In either case, there is a plausible explanation for the tension that many have felt between 7:14 and 7:15.

---

corpus of ancient Hebrew literature that has come down to us in the Bible exhibits a remarkable density of such allusions. Now some may object that the sort of dynamic that comes into play when, say, T.S. Eliot alludes to Shakespeare and Milton cannot be applied to the Bible, which represents a "scribal culture" that makes frequent use not of literary allusion but of traditional formulas, verbal stereotypes. The whole notion of formula, so often invoked in biblical scholarship, needs serious critical re-examination because there is such an abundance of subtle, significant *variations* in the biblical use of formulas. . . . In any case, the Bible offers rich and varied evidence of the most purposeful literary allusions—not the recurrence of fixed formula or conventional stereotype but a pointed activation of one text by another. . . .

[135] Sommer, *Allusion*, 22, 32.

[136] If this is correct, it is another argument against the prevalent view of 9:11 as a late interpolation; see Paul, *Amos*, 288-91, Hayes, *Amos*, 223-27, and the literature cited there. For more on allusion and the criteria for distinguishing it from accidental similarity, see R. Klapper, G. Posner, and M. Friedman, "Amnon and Tamar: A Case Study in Allusions," *Nahalah: Yeshiva University Journal for the Study of Bible* 1 (1999) 23-33 and the literature cited there.

[137] With this interpretation, we take Ben-Porat's third step in actualizing an allusion: "the modification of the interpretation of the sign in the alluding text"; Sommer, *Allusion*, 12.

CHAPTER 5

Amos' Occupations

The Herdsmen from Tekoa

We saw in the preceding chapter that Neh 3:5 appears to contain a reference to the herdsmen of Tekoa. The reappearance of this group centuries after Amos' time supports a theory proposed by S. R. Driver. Driver suggested that the settlement of נקדים at Tekoa may have consisted of "families following hereditary trades," and he compared the "families of scribes dwelling in Jabez" (1 Chr 2:55) and the "families of the linen factory at Beth-Ashbea" (1 Chr 4:21).[1] At Ur, "many of the men appearing in these [Old Babylonian herding] documents are related to each other as brothers and the profession was often passed on from father to son."[2] At Nuzi, too, "the herding profession was hereditary, and in a number of cases families of herdsmen can be found working for the same livestock owner or his family."[3] This line of interpretation would imply that Amos was working in a family business.

One might go a step further and consider the possibility that Tekoa was home to a number of families of herdsmen that worked together

[1] Driver, *Books*, 128.
[2] Van De Mieroop, "Sheep and Goat," 169.
[3] Morrison, "Evidence," 262.

in some sort of professional organization. Various terms have been applied to ancient organizations of this type. G. Alon uses the term "cooperative" to describe the חרמי טיבריה "fishermen of Tiberias," who are mentioned together with the גרוסי צפורין "gristmakers of Sepphoris" and the דשושי עכו "wheat-stampers of Acre" in the Palestinian Talmud (*y. Pes.* 4.1, 30d; *Moed Q.* 2.5, 81b).[4] I. Mendelsohn and S. Appelbaum use the term "guild" to refer to the organizations of wool producers, dyers, bakers, donkey-drivers and ship-owners whose rights are set forth in *t. B. Meṣ.* 11.24-26.[5] Like the "fishermen of Tiberias" and the "gristmakers of Sepphoris," these organizations were located in a single town. Thus, "the wool producers and the dyers are permitted to say: 'We are partners in buying up whatever [wool and dye] comes to town'" (*t. B. Meṣ.* 11.24).

Organizations of herdsmen have also been discerned. The *nqdm* at Ugarit appear in lists of professional groups that are frequently labeled "guilds."[6] The term "collective" has also been applied:

> One taxation document appears to equate a group of *nqdm* as being equivalent to a village for taxation purposes; this document may indicate that the *nqdm* functioned as a kind of land-holding collective.[7]

We should also mention the hamlet called *Kapru-ša-nāqidāti* "Village of herdsmen" in the Neo-Babylonian period[8] and the town of *Ālu-ša-nāqidāti* "City of herdsmen" in the Neo-Assyrian period.[9] Here too we seem to have groups of נקדים living and working together.

[4] G. Alon, *The Jews in their Land in the Talmudic Age* (Jerusalem: Magnes, 1980-84) 1. 167.

[5] I. Mendelsohn, "Guilds in Ancient Palestine," *BASOR* 80 (December, 1940) 19; Applebaum, "Economic Life," 685. Cf. also D. B. Weisberg, *Guild Structure and Political Allegiance in Early Achaemenid Mesopotamia* (Yale Near Eastern researches 1; New Haven: Yale University Press, 1967).

[6] See, for example, Kapelrud, *Central Ideas*, 6; Craigie, "Amos," 31; Cutler and Macdonald, "UT 113."

[7] Craigie, "Amos," 32.

[8] San Nicolò, "Viehwirtschaft I," 284. For the location, see R. Zadok, *Geographical Names According to New- and Late-Babylonian Texts* (Wiesbaden: L. Reichert, 1985) 194. I am indebted to J. A. Brinkman for this reference.

[9] R. Zadok, "Zur Geographie Babyloniens während des sargonidischen, chaldäischen, achämenidischen und hellenistischen Zeitalters," *WO* 16 (1985) 44. The latter town was located in Rashi, a small country sandwiched between Babylonia and Elam. For the Βουκόλων πόλις "City of cowherds" of Roman Palestine, see below.

The "herdsmen from Tekoa" have also been viewed in this way. In Engnell's view, they constituted "a local shepherds' collective subordinate to Jerusalem's temple."[10] Danell, with *Kapru-ša-nāqidāti* in mind, writes that the expression in Amos 1:1 "can plausibly be understood as a designation of a *convivium* or village of tenders of livestock in Tekoa. Based on the definite form, it must have been a commonly known group for that time. . . ."[11] P. de Robert believes that they formed "une sorte de corporation, peut-être celle des propriétaires-éleveurs."[12] Rosenbaum suggests that the נקדים of Amos 1:1 were "perhaps a guild."[13]

It is tempting to draw a further analogy between Tekoa and *Kapru-ša-nāqidāti* in support of one aspect of Engnell's view. The latter village was located in the vicinity of Uruk, and its inhabitants were in charge of flocks and herds belonging to Eanna, the main temple of Uruk.[14] Did the herdsmen of Tekoa have a similar relationship to the Temple of Jerusalem, only 16 km. away? To a certain extent, the analogy is supported by postbiblical sources. According to a *baraita* in the Talmud (*b. Men.* 87a and *b. Sot.* 34b), sheep for the public sacrifices were brought from Hebron.[15] Since Tekoa is situated roughly midway between Hebron and Jerusalem, it is entirely possible that the "herdsmen from Tekoa" were regular suppliers of sheep for the Temple.

There is, then, no problem with the claim of Kapelrud, based on Engnell's theory, that Amos "furnish[ed the cult] with the necessary sheep for sacrifices"; however, it does not follow that "a *noqed* in Judah in the time of Amos . . . may have been a person of rather high rank who was responsible for a large part of the temple herds," let alone that Amos "may thus officially have had something to do with the cult." Andersen and Freedman note correctly that "there are no examples within the OT of any Israelite shrines having their own flocks and shepherds."[16]

[10] Engnell, "Profetismens ursprung," 15.

[11] Danell, *Amos*, 9.

[12] Robert, *Le berger*, 26.

[13] Rosenbaum, *Amos*, 46.

[14] San Nicolò, "Viehwirtschaft I," 284.

[15] Safrai, "The Temple and the Divine Service," 321; *idem*, "The Temple," 882. In the Middle Ages, there were plantations of sycomores around Hebron, according to al-Iṣṭaḫrī and al-Idrīsī; see Goor, "History," 132.

[16] Andersen and Freedman, *Amos*, 188. So too Wright ("Liver," 10): "In contrast to Mesopotamia, there is little evidence that cultic centres kept herds of (*sic*) flocks." It is

In rabbinic literature, too, there is not a single mention of temple flocks and herds or temple herdsmen.[17] The same goes for the writings of Josephus. Both bodies of literature attest to the *purchase* of animals for communal sacrifice, but they are silent concerning the *breeding* and *raising* of such animals.[18] In Roman times, animals for the public sacrifices were purchased by the Temple with funds from the "sacred treasury known as 'Corbonas.'"[19] These purchases could not have been meant to augment existing flocks and herds or start new ones, for there was a requirement that the animals offered as public sacrifices in any given fiscal year be purchased with coins earmarked for that year. During the month of Adar, the Temple began to solicit the donation of "new shekels" (תקלין חדתין) for the following fiscal year, which began on the first of Nisan; from that day on, only animals purchased with "new shekels" could be used for public sacrifice.[20] As for the frequency of these purchases, our only hint is the Mishnah's report (*m. Arak.* 2.5) that at least five inspected lambs were kept on hand in the Lamb Chamber at all times.

The abovementioned evidence for the *time* of purchase is consistent with the evidence for the *place* of purchase. According to *y. Sheq.* 8.2,

---

thus surprising that Wright ("Liver," 11) leans towards the view that "Amos was attached to a cult." Hasel (*Amos*, 37) credits Murtonen ("Hepatoscoper?" 170-71) with showing that there is not "any evidence that the Jerusalem temple or any Israelite shrine ever had such temple flocks and shepherds," but Murtonen is completely silent about this issue. Nor is there reason to believe that the Temple possessed grazing lands. R. de Vaux (*Ancient Israel* [New York: McGraw-Hill, 1961] 379) writes: "Unlike the temples of Mesopotamia and of Egypt, the Temple in Jerusalem did not possess vast tracts of real estate." E. Bickerman's discussion of the Second Temple (*From Ezra to the Last of the Maccabees* [New York: Schocken, 1962] 14) goes further: "While in Egypt [in the fourth century B.C.E.], a very large part of the soil belonged to the temples . . . , the sanctuary of Jerusalem does not appear to have possessed any real estate outside its own site. . . ." For arguments to the contrary, see J. Blenkinsopp, "Did the Second Jerusalemite Temple Possess Land?" *Transeuphratène* 21 (2001) 61-68, and J. Weinberg, *The Citizen-Temple Community* (JSOTSup 151; Sheffield: Sheffield Academic Press, 1992) 95-97.

[17] Here too there is no evidence for temple fields; Safrai, "The Temple," 317.

[18] The closest thing to an exception that I know of is a prohibition on buying the red heifer as a calf and raising it; see *Sifre Zuṭa* to Num 19:2 in ספרי דבי רב (ed. H. S. Horovitz; Leipzig: G. Fock, 1917) 300 lines 14-17.

[19] Josephus, *A.J.* 3.10.1 §237 and *J. W.* 2.9.1 §175; *m. Sheq.* 4.1. For the financing in earlier times, see Y. Liver, פרשת מחצית השקל, in ספר היובל ליחזקאל קויפמן (ed. M. Haran; Jerusalem: Magnes, 1961) 54-67; J. A. Goldstein, *II Maccabees* (AB 41A; Garden City, N.Y.: Doubleday, 1983) 200-201.

[20] See *m. Sheq.* 1.1, 4.1, 6.5 and *t. Rosh ha-Sh.* 1.1, 1.4.

50c, coins found on the Temple mount are presumed to be coins from the treasury that were used to purchase animals for communal sacrifice. This implies that the purchase of animals took place there—not in places like Hebron, Sharon, and Moab, where the animals were raised.

In short, the available evidence supports the view of S. Safrai that "it was the duty of the Temple treasurers to supply the Temple with . . . communal sacrifices."[21] The treasurers purchased these animals in limited quantities, as needed, from suppliers who brought the animals to the Temple. These suppliers were not temple personnel; indeed, they did not even have to be Jews (*t. Sheq.* 1.7). Like the suppliers of flour, wine, and oil, they were private businessmen.[22] They may well have been the same dealers (סוחרי בהמה) who sold animals in the Jerusalem market for private sacrifice (*m. Sheq.* 7.2).[23] In S. Applebaum's words, "the need of sacrifices must have been a permanent incentive to run cattle and sheep for sale in Jerusalem."[24]

Even in Mesopotamia, where the temples *did* own flocks and herds, the people who managed them were not always temple personnel, let alone cult personnel. We have already seen that this is true of the Eanna temple of Uruk in the Neo-Babylonian period: "Especially in Uruk it is obvious that (some? of) the external herding was contracted out. . . ."[25] It is equally true of the Nanna-Ningal temple of Ur in the Old Babylonian period:

> The enormous herds belonging to the Nanna-Ningal temple complex were not herded by temple dependents, but were assigned to private shepherds who combined the care of their own herds with that of the temple animals.[26]

[21] Safrai, "The Temple," 881.

[22] It is clear from *m. Sheq.* 4.9 and *t. Sheq.* 2.11-13 that the Temple's suppliers of flour, wine and oil were private businessmen, paid monthly, who had to agree to bear all losses resulting from spillage, spoilage, and market fluctuations up to the moment when their goods were actually used. Suppliers who guaranteed "price protection" (as it is called today) can hardly have been temple personnel.

[23] Safrai, "The Temple," 320. For the number of lambs sacrificed each year on Passover alone during the Roman period and the size of the national herd needed to produce that number, see J. Pastor, *Land and Economy in Ancient Palestine* (London: Routledge, 1997) 10, 178.

[24] Applebaum, "Economic Life," 655.

[25] Driel, "Sheep and Goats," 224.

[26] Van De Mieroop, "Sheep and Goat," 166-67.

> In the archive of Apil-kittim . . . dating to the years Rīm-Sîn 31-36 (1792-1787), we see how a private businessman and his associates took care of all the daily worries with regard to the temple cattle. The temple is only mentioned as the owner, but the products are managed by Apil-kittim. The temple gave the right to manage the herds and to convert their products into silver to private businessmen, who were allowed to keep any profits they made during these transactions.[27]

The Neo-Babylonian temples were not self-sufficient. They were forced to purchase sheep and cattle "to fill the gap between production and requirements."[28] This was particularly true of the Ebabbar temple in Neo-Babylonian Sippar:

> Especially in Sippar much cattle had to be bought for offerings. This is a clear indication of the existence of cattle breeding outside the institutions, producing for their needs. This strongly suggests the existence of a private sector.[29]

Clearly, we have come a long way since the time when it was possible to assume that "in antiquity, all economic and political life was centralized in the sanctuaries" and to base conclusions about Amos on that assumption.[30] Nowadays, it seems more natural to assume that the herdsmen from Tekoa were self-employed.

The existence of an organization of self-employed stock breeders in eighth-century Israel would seem to represent a substantial shift away from the integration of animal husbandry and agriculture in the subsistence economy of the Highlands in the early Iron Age.[31] In the affluent society described by Amos and Hosea, the demand for meat could no longer be satisfied by small family farms devoted mainly to agriculture.[32] In the words of D. C. Hopkins: "In an agricultural system where fodder production was probably not achieved on any great scale . . . , there are obvious limits to the community's involvement in pas-

[27] Van De Mieroop, "Sheep and Goat," 170.

[28] Driel, "Cattle," 217.

[29] Driel, "Cattle," 233.

[30] Engnell, "Amos," 59-60.

[31] See D. C. Hopkins, *The Highlands of Canaan: Agricultural Life in the Early Iron Age* (Social world of biblical antiquity series 3; Sheffield: Almond, 1985) 245-50.

[32] For meat consumption in eighth-century Israel, see M. Silver, *Prophets and Markets: The Political Economy of Ancient Israel* (Boston: Kluwer-Nijhoff, 1983) 97-98.

toralism beyond which the integration of the two modes of production fractures and specialist stock breeders take to distant pastures."[33] Hence the rise of the herdsmen from Tekoa and of the term נקדים used to refer to them.[34]

The Location of Amos' Sycomores

It has often been noted that Tekoa is too high above sea-level for the sycomore, a tropical tree that cannot tolerate cold. Indeed, there were no sycomores in the wilderness around Tekoa in Jerome's time, a fact that led him to question the Septuagint's rendering of שקמים.[35] The same problem has led others to question the traditional identification of Tekoa.[36] Neither of these responses to the problem is warranted.

Where, then, were Amos' sycomores located? The best-known answer to this question is found in the Targum to 7:14: ושקמין לי בשפלתא "and I have sycomores in the Shephelah" (cf. *b. Ned.* 38a). This paraphrase must have been suggested by the biblical verses that speak of the sycomores of the Shephelah.[37] A number of modern scholars have adopted this view.[38] According to Breslavy, the distance between Tekoa and the Shephelah is no obstacle:

> Even if he did not leave Tekoa, his birthplace, he could have been a sycomore owner far from his city. From Arab agriculture in recent generations, we know that Arabs residing in the mountains had land in the Shephelah and the valleys.[39]

[33] Hopkins, *Highlands*, 248.

[34] It has been suggested that נוקד is a loanword from Akkadian; see chapter 4 at n. 44 above.

[35] Jerome, *Commentarii*, 324 lines 404-7; cf. line 385.

[36] See the literature cited by Weiss (ספר עמוס, 2. 8-9 n. 80), Watts (*Vision*, 34 n. 28), and Rosenbaum (*Amos*, 29-40), and add the refutation of Weippert (*Amos*, 3). Weiss' assertion that Abarbanel accepts the view of David Qimḥi that Tekoa was in the territory of Asher is not completely accurate. A correction in the autograph of Abarbanel's commentary (*Comentario*, 18 line 12) shows that he changed his mind on this issue. He originally wrote בנוקדים אשר בעיר תקוע שהיא עיר אחת מנחלת בני אשר, but he subsequently crossed out the words שהיא עיר אחת מנחלת בני אשר. This correction must be understood in the light of his commentary to 2 Sam 14:2, where he states that Profiat Duran has refuted Qimḥi's opinion on the matter; see Profiat Duran, אגרות ר׳ פריפוט דוראן in ספר מעשה אפֹד (ed. J. Friedländer and J. Kohn; Vienna: J. Holzwarth, 1865) 199.

[37] 1 Kgs 10:27, 2 Chr 1:15, 9:27. See chapter 1 nn. 133-34 above.

[38] See Weiss, ספר עמוס, 2. 446 n. 177.

[39] Breslavy, עמוס, 100.

It is true that if Amos actually *owned* the trees, the distance between Tekoa and the Shephelah would not have been a problem for him. He could have sold the fig harvest in advance to a *gemamzi* in the Shephelah and kept the number of trips from Tekoa to a minimum. However, it seems unlikely that Amos owned the trees. The owner of sycomore trees who wished to impress others would identify himself as a harvester of beams, not figs, since the beams of the sycomore were far more valuable than the figs.[40] It is more likely that he himself was a *gemamzi* and that what he owned was the figs, not the trees.[41]

As noted by Danell, another possibility is suggested by the fact that many of the sycomore trees of the Shephelah were in groves belonging to the crown: "either *bōlēs šiqmīm* gives expression to Amos' well-off status as an owner of mulberry(-fig) plantations or else it implies that he was a worker in the royal mulberry(-fig) plantations in Šefela (1 Kgs

[40] As noted in chapter 1 above, the value of sycomore beams came from their use in the construction of ceilings. According to a tradition attributed to Abba Saul in *t. Men.* 13.20, they were valuable enough to rob: אבא שאול אומר, קורות שיקמה היו ביריחו והיו בעלי אגרוף באין ונוטלין אותן בזרוע. עמדו בעלים והקדישום לשמים "Abba Saul says: There were sycomore beams in Jericho, and thugs used to come and take them by force, so the owners went and consecrated them (the beams) to Heaven." The stump that produced the beams was valuable enough that, when the field in which it stood was sold, it was not included in the sale unless it was explicitly mentioned (*m. B. Bat.* 4.9). In Mesopotamia, too, "the significant and valuable part of a house was its wooden roof-beams"; E. Stone, "Texts, Architecture and Ethnographic Analogy: Patterns of Residence in Old Babylonian Nippur," *Iraq* 43 (1981) 20 apud Moorey, *Mesopotamian Materials*, 355. By contrast, the figs had little value; see chapter 1 nn. 53, 113, 119 and chapter 2 n. 32 above. In this connection too, the authority of Abba Saul has been invoked. According to Rashi's reading of *m. B. Bat.* 2.13 (*b. B. Bat.* 27b), Abba Saul classified the שקמה as a barren tree (אילן סרק) rather than a food tree (אילן מאכל, the biblical עץ מאכל), which fits well with Abba Saul's other statement about the שקמה. Despite the fact that other commentators (e.g., pseudo-Gershom and Maimonides) disagree with Rashi's reading of Abba Saul's statement and the fact that other rabbinic sources (*y. Orlah* 1.1, 70c and probably *Sifra* 90a to Lev 19:23) imply that that the שקמה was classified as a food tree, Rashi's reading is accepted by I. Lewy (*Ueber einige Fragmente aus der Mischna des Abba Saul*, 1876, *apud* Löw, *Flora*, 1. 278, 2. 401) and by Lieberman (תוספתא כפשוטו, 2. 361 n. 28).

[41] For the Egyptian *gemamzi*, see chapter 2 above. The Yemeni *miballis* seems to own the trees as well as the figs; see I. Al-Akwaʿ, *Al-ʾamṯāl al-yamāniyya* (n.p.: Dār al-Maʿārif, 1968) 312 no. 909: "The *muballis* who picks *balas*, that is figs (*tīn*), from his trees and brings them early in the morning to the market to sell them. . . ."

27:28)."[42] Danell also notes the possibility that Amos dealt with "the royal temple herds."[43]

A similar suggestion was made independently by H. Cazelles:

> . . . les termes qui décrivent sa profession *bôqer*, *bôles* (VII, 14) font croire qu'il n'était pas un petit pâtre, mais un fonctionnaire d'Ozias-Azarias, dont on nous dit (II Chr., XXVI 10) qu'il aimait l'agriculture.[44]

This brief remark is puzzling. It is difficult to see how the terms בוקר and בולס provide any support to the claim that Amos was a government official during Uzziah's reign.

Even more perplexing is the use of 2 Chr 26:10: ויבן מגדלים במדבר ויחצב ברות רבים כי מקנה רב היה לו ובשפלה ובמישור אכרים וכרמים בהרים ובכרמל כי אהב אדמה היה. One would have expected Cazelles to focus on the reference to livestock and the Shephelah instead of on the reference to Uzziah being a "lover of the soil." Unfortunately, the syntax of the verse is unclear precisely at that point. If ובשפלה ובמישור modifies מקנה רב היה לו ("He built towers in the wilderness and hewed out many cisterns, for he had much livestock [there] and in the Shephelah and on the plain . . ."),[45] as the Masoretic accents suggest, it would seem that Uzziah had livestock in the Shephelah, in the vicinity of the sycomore groves he inherited from David. If so, Amos could indeed have worked for the king on both of these. However, some modern scholars construe the syntax of the verse differently. Thus, S. Japhet writes:

> The precise division of v. 10 is not entirely clear (cf. also NEB and JPS); it seems that the *Ethnah* accent should be moved for a reading: "and farmers in the Shephelah and the plain" (thus JPS).[46]

The possibility that Amos was a government worker has been raised by several other scholars. After discussing the list of the stewards of

[42] Danell, *Amos*, 10-11.

[43] Danell, *Amos*, 11.

[44] H. Cazelles, "Mari et l'Ancien Testament," in *La Civilisation de Mari: XVe Rencontre assyriologique international . . . 1966* (ed. J.-R. Kupper; Paris: Les Belles Lettres, 1967) 90.

[45] Alternatively (with most translations): "He built towers in the wilderness; and he hewed out many cisterns, for he had much livestock, both in the Shephelah and in the plain. . . ."

[46] Japhet, *I & II Chronicles*, 881.

David's property (1 Chr 27:25-31), which included sycomore plantations and herds and flocks, T. N. D. Mettinger writes:

> In this context I would like to put forward a surmise concerning the profession of Amos. The different allusions made to this (Amos 1,1; 7,14; cf. 7,1) find a common denominator in the royal estate. Besides, Tekoa is mentioned among the royal fortresses (2 Ch 11,6). . . . It is therefore possible that Amos served in this branch of administration before his prophetic activity.[47]

R. R. Wilson points to the term נקדים and its Ugaritic cognate as evidence "that Amos may have been a government employee who was responsible for a fairly sizable herd of sheep, or, alternatively, that he was an independent sheep owner with a large herd."[48] The Ugaritic evidence gains added significance from Craigie's subsequent finding that "the majority of the evidence indicates that the *nqdm* [at Ugarit] were servants of the royal establishment."[49]

Such arguments are intriguing but far from conclusive. The other alternative raised by Danell and Wilson certainly represents the consensus of modern scholarship. The evidence presented in this monograph does nothing to undermine that consensus. It suggests that Amos was self-employed and that he owned both the livestock that he tended and the sycomore figs that he harvested (but not the trees themselves).

In any event, most scholars look for Amos' sycomores closer to Tekoa.[50] In 1928, W. Rudolph visited Tekoa and found that:

> . . . die Bewohner des Ortes ihre Ackerfelder und Baumanlagen in den tiefer gelegenen, oft ziemlich weit entfernten Mulden und Tälern hatten, wo es auch Sykomoren gibt. Man braucht also gar nicht Weidegänge bis in die Schefela anzunehmen. . . .[51]

[47] Mettinger, *State Officials*, 87-88.

[48] R. R. Wilson, *Prophecy and Society in Ancient Israel* (Philadelphia: Fortress, 1980) 268. His first alternative is accepted by Rosenbaum (*Amos*, 46), who assigns Amos "a middle-level position in his government's service."

[49] Craigie, "Amos," 33.

[50] See Weiss, ספר עמוס, 2. 446 n. 177.

[51] Rudolph, *Joel—Amos—Obadja—Jona*, 258. At the beginning of the century, Masterman ("Sycomore," 2877) found something similar for the village of Silwan.

Several scholars locate Amos' sycomores near the Dead Sea.[52] Others look to the Jericho Valley.[53] Both views are possible, but the latter has more evidence in its favor. The Jericho Valley was known for its sycomores in Roman times,[54] and it is not impossible that some of these trees had survived from Amos' time.[55] The beams cut by Elisha's disciples at the Jordan River (2 Kgs 6:2-5) may well have been from sycomores[56] growing near Jericho.[57] H. B. Tristram reports finding "a few gnarled and aged sycomores among the ruins by the wayside at ancient Jericho, and by the channel of the Wady Kelt," and Y. Feliks describes a giant sycomore alive in Jericho today.[58] We shall return to this question below.

Linking the Two Occupations

Is there a link between Amos' two occupations? The best answers to this question have been given by exegetes who assumed that Amos had

[52] Driver, *Books,* 212; R. S. Cripps, *A Critical and Exegetical Commentary on the Book of Amos* (2nd ed.; London: SPCK, 1955) 10; Hammershaimb, *Amos,* 118; Y. Feliks, טבע וארץ בתנ"ך (Jerusalem: R. Mass, 1992) 143 n. 10. (I am indebted to M. Jacobowitz for the last reference.) Most of these are quoted below.

[53] B. E. Willoughby, "Amos, Book of," in *ABD* 1. 203; N. Hareuveni, *Tree and Shrub*, 91 (see below).

[54] See *t. Men.* 13.20 (quoted in n. 40 above), *m. Pes.* 4.8, and Luke 19:1-4.

[55] For the life-span of this tree, see Galil, השקמה, 313: "The exact age of the old trees has not been established, but an age of 500-800 is apparently commonplace." For the rabbinic estimate of 600 years, see Galil *ad loc.* and N. Hareuveni, *Tree and Shrub*, 88. The "cathedral fig tree" in Famagusta is said to be the oldest living thing in Cyprus, having been planted in front of St. Nicholas Cathedral while it was being built (1298-1312 CE). Kruger National Park in South Africa contains a sycomore said to be well over 1,000 years old. Finally, Y. Shapira (גילן של שקמים עתיקות, *Teva Vaaretz* 9 [1967] 28-29) argues that one modern specimen in Israel must be several centuries older than the Byzantine pool that was constructed next to it after its roots were exposed by erosion.

[56] Rabbinic literature knows of two principle sources of construction beams: the cedar of Lebanon and the sycomore. Cf. Isa 9:9, discussed in chapter 1 n. 120 above.

[57] The story suggests that the school that they had outgrown was not very far from the Jordan. This makes sense according to 2 Kgs 2:18-22, which is generally taken to mean that Elisha settled in Jericho after accompanying Elijah to ערבות מואב על ירדן ירחו, only seven kilometers away. Indeed, the site selected for the new school may have been the very spot where Elijah and Elisha crossed the Jordan. As noted by Abarbanel (in his commentary to 2 Kgs 6:2), that would be the perfect spot for training would-be prophets.

[58] Tristram, *Natural History*, 399; Feliks, עצי־פרי, 167 n. 48.

only one occupation, that of a herdsman, because these exegetes were forced to ponder the benefits that the sycomore could provide to the animals in Amos' care.

For Eliezer of Beaugency, the major benefit was shade. According to him, the context suggests that בולס שקמים somehow refers to pasturing in the shade of sycomore trees.[59] Shade, in fact, is mentioned by many modern students of the sycomore as one of its salient characteristics.[60] The very name of the sycomore in Egyptian, *nh.t*, alludes to this characteristic.[61] That sycomores (and carobs) provide more shade than other trees seems to be implicit in *m. B. Bat.* 2.13.[62] The same shade that is bad for crops is good for livestock.

For Sherira Gaon, as cited by Ibn Janāḥ, the benefit was nourishment, and בולס שקמים refers to mixing sycomore leaves into fodder.[63] This view has reached modern scholars fourth- and fifthhand through M. Bič and T. J. Wright, who learned of it from Ibn Parḥon's Hebrew abridgment of Ibn Janāḥ's dictionary.[64] It should be noted that Ibn Parḥon substitutes "barley" for Ibn Janāḥ's "fodder." Wright asks two questions, only one of which need detain us:

[59] Eliezer of Beaugency, *Kommentar zu den XII kleinen Propheten*, 2. 52.

[60] See Figari, *Studii*, 177; Goldmann, *La figue*, 46; R. Muschler, *A Manual Flora of Egypt* (Berlin: R. Friedlaender & Sohn, 1912) 249; H. N. Moldenke and A. L. Moldenke, *Plants of the Bible* (Waltham, Mass.: Chronica Botanica, 1952) 107; Wright, "Sycomore Fig," 365; Kislev, השקמים, 24.

[61] Baum, *Arbres*, 36, citing H. Fischer, "Another example of the verb *nh* 'shelter,'" *JEA* 64 (1978) 131-32.

[62] This mishnah allows the owner of a field to cut off *all* of the branches of a sycomore or carob tree in a neighboring field that overhang his property, because their rich foliage blocks the sunlight and harms the field. For other trees, one may cut off only branches low enough to interfere with plowing. In modern Ethiopia, too, the sycomore is lopped to reduce shade; Bekele-Tesemma, *Useful Trees*, 250.

[63] Ibn Janāḥ, *ʾUṣūl*, 96 lines 5-9. For Ibn Janāḥ's citations of Sherira Gaon, see S. Abramson, מפי בעלי לשונות (Jerusalem: Mossad Harav Kook, 1988) 290-92. Sherira connects בולס שקמים with Mishnaic Hebrew עיסה בלוסה "mixed dough," i. e., dough from flour mixed with bran. David Qimḥi, followed by Abarbanel, combines Sherira's interpretation of בולס שקמים with the one-occupation view: "a gatherer (לוקט) of sycomore(-fig)s for his cattle to eat."

[64] Bič, *Amos*, 156; Wright, "Sycomore Fig," 366. See Ibn Parḥon, מחברת הערוך, 2. 9b. Curiously, Shelomoh Ibn Parḥon goes by the German (< Vulgate < LXX) name "Salomon," instead of "Solomon," in English discussions of our verse; see Wright, "Sycomore Fig," 366; Hayes, *Amos*, 236; Paul, *Amos*, 248. That is no doubt because Wright followed Bič in citing the text from a German article by Bacher but failed to substitute the English equivalent of שלמה. Even more curious is Danell's attribution

> The second question—whether livestock ate a mixture of barley and sycomore leaves—is something I am unable to ascertain from other sources. . . .
>
> Instead of sycomore leaves, it is just possible that the fruit of the fig would be used as fodder, especially those which are not induced to ripeness by wounding or gashing. . . . Such fruit are full of dead, male and some female, wasps. They are not palatable, and as such "at best, they are used by poor farmers and by bedouins as fodder for goats and other domestic animals." Perhaps the concern of Amos was simply to provide fodder for those in his charge.[65]

Uncertainty about the use of sycomore leaves as fodder was expressed already by Gesenius: "Sed folia sycomori armentorum pabulum fuisse, aliunde non constat."[66]

Wright's revision of Sherira's interpretation from leaves as fodder to figs as fodder is also not new. According to Bochart's plausible reading of David Qimḥi's commentary to Amos 7:14,[67] the same revision is found there. The idea that Amos used sycomore fruit as fodder for his herd would later be put forward by G. Dalman and J. A. Soggin ("during the dry season") as well.[68] Bochart labeled this idea "absurd," on the grounds that sycomore fruit is not food for sheep.[69]

There is no justification for either Bochart's doubts about the use of the figs as fodder or the doubts of Gesenius and Wright about the use of the leaves as fodder. The use of sycomore figs as fodder by poor farmers and bedouin shepherds in modern Israel is noted above.[70] In

("Amos," 10) of this interpretation to the third-century Palestinian *amora* Resh Laqish, without supplying a reference. There is clearly some confusion here, since what Danell has translated into Swedish is the comment of the eighteenth-century exegete David Altschuler in his מצודת דוד, printed in the Rabbinic Bible. Danell seems to have interpreted Altschuler's ר״ל "i.e.," as "Resh Laqish."

65 Wright, "Sycomore Fig," 366-67. Cf. n. 32, where he adds that "Roman farmers did use 'fig'-leaves (among others) for fodder when green forage was not available." Note that it is only untreated sycomore figs that are full of wasps and unfit for human consumption. It is an exaggeration to claim that "although the poor did eat sycamore figs, the fruit was mostly used for cattle fodder" (Willoughby, "Amos," 204).

66 Gesenius, *Thesaurus*, 213.

67 Bochart, *Hierozoicon*, 1. 384 (ed. Rosenmüller, 1. 406).

68 G. Dalman, *Arbeit und Sitte in Palästina* (Gütersloh: C. Bertelsmann, 1928-39) 1. 63; Soggin, *Amos*, 10.

69 Bochart, *Hierozoicon*, 1. 384 (ed. Rosenmüller, 1. 406).

70 Wright (above at n. 65) quoting J. Galil and D. Eisikowitch, "Flowering Cycles and Fruit Types of *Ficus Sycomorus* in Israel," *The New Phytologist* 67 (1968) 752.

Egypt, the leaves and fruit of the sycomore serve today as fodder for animals, and the ancient monuments show goats and cattle browsing on the foliage of trees and bushes in semidesert zones.[71] In Ethiopia, "figs [of *F. sycomorus*] are eaten by livestock."[72] In East Africa, the "leaves/figs" of the sycomore "provide livestock forage in the dry season" and are "believed to stimulate milk production in cows."[73] In the Sahel:

> The fruit [of the sycomore] drop when still immature. They are eaten particularly by goats and sheep, but also by cattle and birds. . . . Leaves . . . are a much sought fodder. The tree is therefore lopped.[74]

In the tropical Sudanian and the equatorial Guinean zones, the *Ficus gnaphalocarpa* Steud. (= *Ficus sycomorus* L.) is "consumed by livestock (leaves and fruit)."[75] In South Africa:

> The Bantus feed cows with the foliage and fruit [of the sycomore] to stimulate milk-production. Research in this connection has revealed that the leaves digest easily and have a high nutritional value.[76]

Wright accepts Sherira's assumption that Amos had only one occupation. Indeed, he finds it so convincing that he uses it to critique other interpretations:

> There is one major objection to [the interpretation of the LXX], other than the passage of time, and this is whether Amos would carry out both the task of shepherd and that of carer for sycomores simultaneously, especially in the light of the division of tasks as illustrated in 1 Ch. xxvii 25-31.[77]

[71] Baum, *Arbres*, 211.

[72] Bekele-Tesemma, *Useful Trees*, 250.

[73] Food and Agriculture Organization, *Food Plants*, 290.

[74] Maydell, *Trees*, 273.

[75] H. N. Le Houérou, "The Role of Browse in the Sahelian and Sudanian Zones," *Browse in Africa: The Current State of Knowledge* (Addis Ababa: International Livestock Centre for Africa, 1980) 91. For the term *Ficus gnaphalocarpa* Steud., see Introduction n. 6 above.

[76] P. Van Wyck, *Trees of the Kruger National Park* (Cape Town: Purnell, 1972-74) 1. 65. For the results of a chemical analysis of the leaves, see F. Busson, *Plantes alimentaires de l'ouest africain: étude botanique, biologique et chimique* (Marseille: L'Imprimerie Leconte, 1965) 107-9.

[77] Wright, "Sycomore Fig," 368. Cf. already Rudolph, *Joel—Amos—Obadja—Jona*, 257 n. 18.

Wright's argument for the one-occupation theory based on the structure of the royal bureaucracy described in 1 Chr 27:25-31 is not convincing. There is no reason to expect a small business to exhibit the same degree of specialization as the royal bureaucracy. Moreover, Wright fails to explain why Amos would feel the need to tell Amaziah how he feeds his animals.

The usual assumption among modern scholars is that Amos was able to juggle two jobs. In support of this assumption, A. S. Yahudah is often cited:

> En Orient les sycomores poussent très souvent près des puits. Les bergers s'occupent de l'incision des fruits pendant que leurs troupeaux paissent ou s'abreuvent. בולס שקמים n'est pas un métier à part, mais ce peut être l'occupation accessoire d'un berger. . . .[78]

In summary, the sycomore tree provides both shade and food for livestock, and it provides a good vantage point for the herdsmen to keep an eye on them. Its dependence on large amounts of water assures that there is always a nearby source of water for the animals to drink. Amos—or his children or employees or partners—could easily have tended animals and sycomores at the same time. This is particularly true if the animals were bovids, since "cattle require less labor to control and maintain than sheep-goats."[79] All of this calls to mind Strabo's description of some ruined cities on the coast of Palestine south of Mt. Carmel. Immediately after Συκαμίνων πόλις "City of sycomores," Strabo mentions Βουκόλων πόλις "City of cowherds"![80] Could this be more than a coincidence? Is it possible that cowherds were drawn to the area by the sycomores?

The two professions are, thus, quite compatible, but even if they were less compatible, that would not be grounds for rejecting the two-occupation view. We have already cited van Driel's conclusion that "in Uruk the *nāqidu* was a person with sometimes wide ranging interests

[78] A. S. Yahuda *apud* Goldmann, *La figue*, 45 n.1.

[79] R. W. Redding, "Subsistence Security as a Selective Pressure Favoring Increasing Cultural Complexity," *Bulletin on Sumerian Agriculture* 7 (1993) 86.

[80] Strabo, *Geography* 7. 274-75 (16.2.27). For the use of βούκολος by Aquila, Symmachus and Theodotion to render בוקר in Amos 7:14, see chapter 4 above. For Συκαμίνων πόλις, see chapter 1 above.

in cattle and sheep herding and in arable farming."[81] Van Driel points to Iqišā son of Nannā-ereš, the *nāqidu* in charge of 500 cows and 2,000 sheep and goats discussed above, adducing evidence that he also raised barley.[82] Although barley was used as fodder, especially in the winter,[83] it is arguable that Iqišā's combination of occupations is less natural than Amos'. In any event, Iqišā shows that there is no problem in assuming that Amos was a *nāqidu ša ṣēni u lâti* who did a little farming on the side.

A number of scholars have attempted to find a suitable locale for this combination. In S. R. Driver's view:

> We must suppose the "naḳad-keepers of Tekoa" (i.1) to have owned lands in the "wilderness" or pasture-ground, stretching down to the Dead Sea on the east . . . ; and here, in some sufficiently sheltered situation, must have grown the sycomore trees, which the prophet "dressed."[84]

According to Hammershaimb:

> [Sycamores] . . . can also be cultivated in the warm Jordan valley, and in the fertile oases by the Dead Sea. This does not make it impossible that Amos should have supported himself as a sideline by growing sycamores at one of these places, which are near enough to Tekoa for this to be combined with his work as a shepherd. He was probably able to drive his herds with him when he went to attend to his sycamore trees.[85]

Feliks is similar but a bit more specific:

> Amos apparently brought his cattle to the Dead Sea valley, where there are springs such as Ain Feshkha, next to which grow enormous quantities of reeds, which serve as pasturage for cows. Sycomore trees also grew in the area.[86]

N. Hareuveni offers a theory explaining not only where but also how and why the two occupations would have been combined:

[81] Driel, "Sheep and Goats," 225.

[82] See Driel, "Sheep and Goats," 225 and chapter 4 above.

[83] See below.

[84] Driver, *Books*, 212.

[85] Hammershaimb, *Amos,* 118.

[86] Feliks, טבע וארץ, 143 n. 10.

> Amos, who was "among the herdsmen of Tekoa" . . . , must have followed the practice of other shepherds of the area. At the end of the dry, hot summer, when all the pasturage was gone from the Judean Desert, he would move his herds of goats and sheep to the Jordan Plain in the Jericho Valley. This is an area rich in green forage throughout Israel's scorching summer season. . . .
>
> The appropriate season for piercing the sycomore fruit, at least for the sycomores growing in the Jericho Valley, was around the time when the shepherds descended from the desert slopes of Judea and Samaria into the valley. Flocks could graze in the valley, while the shepherds could "moonlight" at other jobs. It is reasonable to assume that the sycomore owners utilized this convenient fact to offer grazing rights in exchange for dressing the sycomore fruit. The shepherd could perch on the sycomore's broad branches and keep a lookout over his flocks while doing the monotonous work of piercing and oiling the still-green fruit. The sycomore owners, on the other hand, were assured of a top-grade crop.[87]

P. J. King considers this "a plausible explanation" of "how Amos could be 'a herdsman, and a dresser of sycamore trees' at the same time."[88] Nevertheless, he questions one aspect of it: "In the opening verse of the Book of Amos he is described as a *noqed*, probably a wealthy landowner and farmer, so Amos may not have been a simple shepherd on hired land."[89] This is not a serious objection, since the term נוקד has nothing to do with land ownership.[90]

The real problem with Hareuveni's theory is that, under the financial arrangement it posits, Amos was not deriving any income directly from tending sycomores. If so, it is not clear why he would mention this activity to Amaziah as evidence of his self-sufficiency. It seems more likely that, like the modern Egyptian *gemamzi*, he bought the fruit harvest in advance.[91] Put differently, the herdsmen from Tekoa

[87] N. Hareuveni, *Tree and Shrub*, 90-91. For "the appropriate season," see below. For "oiling," see chapter 1 n. 55 above. Cf. Willoughby, "Amos," 203: "Amos . . . may also have cut figs in exchange for grazing rights."

[88] P. J. King, *Amos, Hosea, Micah—An Archaeological Commentary* (Philadelphia: The Westminster Press, 1988) 117.

[89] King, *Amos*, 117.

[90] See chapter 4 above.

[91] See chapter 2 above. The sycomore tree produces 3-6 generations of fruit every year. In the Cairo district, only the earlier generations are gashed and eaten; see E. Sickenberger, *Contributions à la Flore d'Égypte* (Mémoires présentés à l'Institut égyptien

were tenants for part of the year. The rent that they paid, perhaps with animals or animal products (wool, milk, etc.), gave them the rights to all vegetation and fruit in the fields that they leased.

In Roman times, the renting of fields containing sycomore trees was common enough to be discussed by the Mishnah (*m. B. Meṣ.* 9.9). The rabbis ruled that only a long-term renter (seven years or more) had the right to harvest the beams of a sycomore. It goes without saying that they placed no such restriction on the figs.

The herdsmen would have sold the good figs and, as Wright suggests, used the inedible figs and the leaves as fodder for their animals. This arrangement was beneficial to the trees as well as the animals:

> The grazing of animals on fallow land, orchard land, and land freshly harvested is particularly important . . . , since the manure deposited by the animals helped to maintain the fertility of the fields.[92]

Hareuveni's assumption that Amos' animals migrated every year to the Jericho Valley is quite reasonable. In this connection, we may cite more fully Hopkins' discussion of the connection between seasonal migration of livestock (transhumance) and specialized stock breeding:

> In an agricultural system where fodder production was probably not achieved on any great scale . . . , there are obvious limits to the community's involvement in pastoralism beyond which the integration of the two modes of production fractures and specialist stock breeders take to distant pastures.[93]
>
> Over time, when the integration of animal husbandry and agriculture would break down and the edge in the competition for resources (fields versus pasture lands) would shift toward the farming sector, the most likely pathway of divorce in the ancient Highlands was some form of

---

4/2; Cairo 1901) 279; Brown and Walsingham, "Sycamore," 10-11. See also chapter 1 n. 53 above. This implies that the *gemamzia* there purchase only the first few crops. If Amos did not bring his herds to the sycomore groves until the end of summer, as Hareuveni assumes, he would have purchased only the *last* few crops.

92 Hopkins, *Highlands*, 247. However, there is little evidence for the manuring of fields in Israel (or in Egypt, for that matter) until the time of the Mishnah; see Y. Feliks, החקלאות בארץ ישראל בימי המקרא המשנה והתלמוד (Jerusalem: R. Mass, 1990) 78-101.

93 Hopkins, *Highlands*, 248.

> transhumance. Ecological conditions permitting the movement of flocks and herds accompanied by some segment of a community or specialist shepherds to seasonal pastures at some remove from the home settlement appear to exist within the Highlands and adjacent areas. . . . In the Judean and Samarian Highlands seasonal migration may have been directed along wadi beds toward the Jordan valley, as has been reported for 19th century Bedouin inhabitants of the Transjordan. . . .[94]

A similar point has been made about stockbreeding in Iraq:

> All sheep-breeding depends on seasonal migration. In winter and spring shepherds follow the growth of vegetation in the steppes and deserts, and retire to the rivers and canals as the summer heat dries it up, pasturing flocks upon empty lands close to the water-courses and marshes or upon cultivated fallows, after making an agreement with the cultivators whereby the fellahin receive a rent in animals or their produce in addition to the manuring of their fields.[95]

At what time of the year did this seasonal migration become necessary in the Tekoa region? According to N. Hareuveni, it was "at the end of the dry, hot summer, when all the pasturage was gone from the Judean Desert." The lack of pasturage lasts well into the winter, and the only alternative to transhumance is supplemental feeding:

> Winter feeding of livestock was a serious perennial problem that faced the stockowner in all Mediterranean lands. To solve it resort was had to a variety of fodder such as straw, branches, young shoots, hay, rice-stalks, unripe corn-stalks, carobs and gourds, and the pods of peas and lentils.[96]

A similar situation obtains today in southern Iraq:

[94] Hopkins, *Highlands*, 250.

[95] Naval Intelligence Division, *Iraq and the Persian Gulf* (Geographical Handbook Series, B.R. 524, 1944), 467, cited by Van De Mieroop, "Sheep and Goat," 170. For further discussion of transhumance in the ancient Near East, see G. M. Schwartz, "Pastoral Nomadism in Ancient Western Asia," in *Civilizations of the Ancient Near East* (ed. J. M. Sasson; New York: Scribner, 1995) 1. 249-58; M. P. Streck, *Das amurritische Onomastikon der altbabylonischen Zeit* (AOAT 271; Münster: Ugarit-Verlag, 2000-) 1. 55-59.

[96] Applebaum, "Economic Life," 656.

> For most of the year sheep secure their sustenance from the grasses and sedges available at pasture but during the winter there is a period of up to four months when little or no pasturage remains and they must be fed grain. . . . An individual sheep without sufficient pasturage is fed two handfuls of barley twice a day.[97]

The problem of feeding livestock in the winter must have been particularly acute for cattle owners, since "cattle require greater access to higher quality forage [than sheep-goats]."[98] Already in the Old Babylonian period we read of this problem: "That straw has been used up and what shall your oxen eat?"; "Since yesterday I have no barley and straw; they are starving."[99] Indeed, for cattle, supplementary feeding was necessary *before* the winter: "Do the oxen that have not eaten fodder in months IV-VI stay alive?"[100]

The sycomore has a number of characteristics that make it uniquely suited for solving—or at least alleviating—this problem. First, much of its fruit, if not gashed and picked in time, becomes infested with wasps and unfit for human consumption. Consequently, a substantial portion of its unusually abundant yield is available for use as fodder. Second, the figs can be stored for several months.[101] The sun-drying of sycomore figs, practiced by Palestinian Arab villagers in modern times,[102] may well go back to antiquity. Third, Tristram reports that, in Palestine, "the Sycomore bears continuously, and I have gathered the figs from November to June."[103] This report may well relate to "the hot Jordan valley" where "there is a tropical temperature throughout the year."[104] Galil, Stein and Horovitz write: "*F. sycomorus* can . . . bridge

97 E.L. Ochsenschlager, "Sheep: Ethnoarchaeology at Al-Hiba," *Bulletin on Sumerian Agriculture* 7 (1993) 33.

98 Redding, "Subsistence," 86.

99 M. Stol, "Old Babylonian Cattle," *Bulletin on Sumerian Agriculture* 8 (1995) 195.

100 *Loc. cit.*

101 Murray, "Fruits," 622.

102 Galil, הפיקוס, 73.

103 Tristram, *Natural History*, 400.

104 Ibid., p. 398. So too in Egypt: "Figs are found upon the trees at all seasons of the year. . . . After the appearance of the third crop there is a continuous but less abundant production of fruit during the autumn and even throughout the winter"; Brown and Walsingham, "Sycamore," 5. On the cooler coastal plain, where Galil made his observations, "the figs are found occasionally on trees even during winter months, but their

over gaps in fruit supply. . . . There is only a short fruitless interval during the coldest months of the year."[105] Fourth, as noted above, the figs and leaves are believed to stimulate milk production in cows.[106] Fifth, its leaves "persist throughout the year, except in the cooler regions, where most of them may fall in winter."[107]

Finally, it should be noted that the sycomore is used to solve a seasonal forage problem in West Africa—not in the winter but in the dry season:

> The young leaves [of *Ficus gnaphalocarpa* Steud.] are consumed by sheep and cattle; during the dry season, the tree is lopped to produce browse for sheep and goats, and sometimes for cattle in very dry seasons, particularly in Ghana.[108]

Our theory, then, is that the herdsmen from Tekoa rented fields containing sycomore trees at the end of summer, when the trees were full of figs. While keeping an eye on their animals, they harvested the figs, selling the edible ones and storing the others. When winter came, they fed their animals the stored figs plus the leaves on the trees and whatever additional figs had appeared on the trees in the meantime.

If the fields they rented were in the Jericho Valley, they were only around 20 km. from Jerusalem—not much further from the city than the spring pastures in the wilderness of Tekoa. If they were in the Jordan Valley, they were even closer. This constant proximity to Jerusalem would have made it easy for Amos and his business associates to supply a steady stream of animals for sacrifice in the Temple throughout the year.

---

number is usually small, and development very slow"; Galil, "Ancient Technique," 180-82. The difference in climate between the Jordan valley and the coastal plain affects the leaves, as well; see n. 107 below.

105 Galil, Stein & Horovitz, "Origin," 197.

106 Food and Agriculture Organization, *Food Plants*, 290; Van Wyck, *Trees*, 1. 65.

107 Hepper, *Pharaoh's Flowers*, 58. Dalman observed sycomores clad with leaves on February 9, 1909 near the Jordan; near the cooler Mediterranean coast, he saw sycomores still nearly bare on April 5, 1921.

108 M. Baumer, *Notes on Trees and Shrubs in Arid and Semi-Arid Regions* (Rome: Food and Agriculture Organization of the United Nations, 1983) 185. For the term *Ficus gnaphalocarpa* Steud., see Introduction n. 6 above.

Amos' Occupations and His Prophecies

Is there a connection between Amos' occupations and his prophecies? This question was well formulated by Abarbanel at the very beginning of his commentary to Amos:

> What is the significance of the phrase אשר היה בנקדים מתקוע? With the rest of the prophets, we do not find that Scripture tells what their profession or occupation was; instead, it is satisfied to mention their names and, at times, to relate them to their ancestors or their land. Why then with Amos . . . does it also mention his profession—that he was one of the נקדים, i. e., a herder of livestock? What does this have to do with his prophecy?[109]

There is in fact a long tradition of reading the book in the light of Amos' life as a herdsman, a tradition summarized by Baur.[110] There has even been an attempt to find parallels between his work with sycomore figs and his work with people.[111] A major component of this tradition in modern times has been the romantic view that the grandeur of nature in the countryside or the solitude of the desert shaped Amos' personality and his Weltanschauung.[112] This view was so axiomatic for J. Morgenstern that he attempted to use it to determine whether Amos was a shepherd or a cowherd:

> . . . unquestionably his regular occupation as a shepherd, with its constant occasion for roaming with his sheep in the solitude of the waste country adjacent to his native Tekoa, far from the settled abodes of men, furnished ample opportunity for quiet meditation and for the visions which he experienced, far better than had he, as a cow-herd, been obliged to remain constantly in close contact with the village and its human relations, an opportunity almost indispensable for the psychic experiences which led to his consciousness and conviction of his prophetic commission.[113]

The degree to which this view is culturally determined can be appreciated by comparing Abarbanel's own answer to his question. It is

[109] Abarbanel, *Comentario*, 2-4.

[110] Baur, *Der Prophet Amos*, 122-25.

[111] S. Bartina, "'Hiendo los higos de los sicomoros' (Am 7, 14)," *EstBib* 25 (1966) 354.

[112] For criticism of this view, see Soggin, *Amos*, 6.

[113] J. Morgenstern, "Amos Studies I," *HUCA* 11 (1936) 36.

based on the *opposite* assumption: Amos became a prophet not because of his occupation but in spite of it. According to him, the point that Scripture wishes to make is that even a lowly, destitute, uneducated herder can be called to prophesy.[114] The training received by the בני הנביאים was not a prerequisite.

Another aspect of Amos' prophecy that has been connected to his life as a herdsman is his imagery. As noted by Baur,[115] it was Jerome who first made this connection in explaining the reference to roaring and shepherds in Amos 1:2:

> It is natural that all professionals speak in examples of their profession and that each one offers a metaphor from the endeavor in which he spends his time. For example, one who is a sailor and a helmsman compares his sadness to a tempest; injury he calls a shipwreck; his enemies he calls head winds; . . . .[116]

In modern times, a similar, but more modest, claim has been made concerning the "cows of Bashan" metaphor in Amos 4:1-3.[117]

Naturally, one must be cautious in making such claims. Amos also employs images from many fields in which he had no special expertise, and it would be arbitrary to focus on the herding images and ignore the others. Nevertheless, there is one verse in which Amos' profession seems to show through with particular force: כאשר יציל הרעה מפי הארי שתי כרעים או בדל אזן כן ינצלו בני ישראל הישבים בשמרון בפאת מטה ובדמשק ערש "as a shepherd rescues from the mouth of a lion two legs or a piece of an ear, so shall the Israelites dwelling in Samaria escape with the end of a bed or the pillow of a couch" (Amos 3:12).[118] It has long been customary to interpret this verse in the light of Exod 22:12: אם טרף יטרף יבאהו עד הטרפה לא ישלם, "if it was torn by beasts, [the shep-

[114] Abarbanel, *Comentario*, 18-20.

[115] Baur, *Der Prophet Amos*, 122 n. 82.

[116] Jerome, *Commentarii*, 215 lines 100-104: "Naturale est, ut omnes artifices suae artis loquantur exemplis, et unusquisque in quo studio triuit aetatem, illius similitudinem proferat: uerbi gratia, qui nauta est et gubernator, tristitiam suam comparat tempestati, damnum, naufragium uocat; inimicos suos, uentos appellat contrarios; . . . ." I am indebted to D. Berger for his help in translating this passage.

[117] T. Kleven, "The Cows of Bashan: A Single Metaphor at Amos 4:1-3," *CBQ* 58 (1996) 215-27, esp. 226.

[118] I hope to defend this translation on another occasion.

herd] shall bring it as evidence; he need not pay compensation for what has been torn by beasts"; and Gen 31:39: טרפה לא הבאתי אליך אנכי אחטנה מידי תבקשנה "that which was torn by beasts I never brought to you; I myself made good the loss; you exacted it from me."[119]

The image is arresting and atypical. Biblical pastoral imagery normally focuses on the *responsibility* of the hired herdsman to safeguard the well-being of the animals in his charge. This is the case whether the image is that of the good shepherd (as in Ezek 34:11-16a and Ps 23:1-4) or the bad one (as in Jer 23:1-2, Ezek 34:1-10 and Zech 11:16). Amos' simile is unique in evoking a *right* of the herdsman, the right to bring evidence absolving him of responsibility for losses. This right is a legal nicety that members of other professions would be unlikely to care about.

The simile is striking in another way. Andersen and Freedman write that "the images are mixed, perhaps incongruous."[120] Even if this incongruity is only apparent,[121] it strengthens the point. The idea of comparing the remnants of a broken luxury couch to the remains of a torn sheep might have been less likely to occur to a person in another business.

Amos' profession may explain some of his imagery, but it certainly does not explain his message. It used to be fashionable to assume that Amos, as a champion of the poor, must himself have been poor—a simple shepherd, who subsisted on a type of fruit that was often fed to animals. That this assumption is no longer in vogue may be due, in part, to the fact that Marxist movements have often been led by wealthy or bourgeois "class traitors." Amos was a traitor not only to his class but also to his collective.[122] The evidence presented above suggests that many of the animals raised by the herdsmen from Tekoa were destined for the altar, at least as private sacrifices. And yet Amos preached: כי אם תעלו לי עלות ומנחתיכם לא ארצה ושלם מריאיכם לא אביט, "When you offer me burnt offerings, I will not accept your gifts, I will

[119] To what extent 1 Sam 17:34-35 is comparable is not clear.

[120] Andersen and Freedman, *Amos*, 409.

[121] I hope to show that this is the case on another occasion.

[122] I leave aside his diatribe against the *marzēaḥ*-banqueters who consume כרים מצאן ועגלים מתוך מרבק "lambs from the צאן and calves from the stall" (Amos 6:4). These are products of the fattener, not the herdsman. For the distinction, see chapter 4 n. 34 above.

pay no heed to your offerings of fatlings" (Amos 5:22).[123] Such talk cannot have been good for business. Financial self-interest would have dictated the opposite message or, at the very least, a discreet silence.

The members of Amos' collective would no doubt have been happier with Malachi's message: והבאתם גזול ואת הפסח ואת החולה והבאתם את המנחה הארצה אותה מידכם "and you bring the stolen, the lame, and the sick—you bring them as a gift; will I accept it from you?" (Mal 1:13). From their point of view, there was a world of difference between Malachi's הארצה and Amos' לא ארצה. The former refers to rejection based on a physical or legal blemish in the offering; the latter, to rejection based on a moral blemish in the offerer. The former would have increased their profits; the latter must have reduced them. Thus, in pursuing the call to prophesy, Amos was not only neglecting his livelihood, he was undermining it as well.

[123] The translation offered here reflects only one of the possible syntactic analyses for the first clause.

Summary

Amos worked with both sycomores and livestock. In Amos' time, the beams of the sycomore were already being used for construction, and they were far more valuable than the figs. Nevertheless, the linguistic evidence does not support Rashi's view that Amos' interest in the sycomore was silvicultural rather than horticultural. Amos worked with the figs, but the precise nature of his work hinges on the meaning of the phrase בולס שקמים, in which the first word is a *hapax*.

Many ancient, medieval and modern readers of Amos, beginning with the Alexandrian translators, have taken בולס as referring to the practice of gashing sycomore figs, a few days before they are picked, to hasten their ripening. This practice has been known in Egypt and Cyprus for thousands of years, and has attracted an enormous amount of attention in travel accounts, pharmacological treatises, ethnobotanical studies, etc. from the time of Theophrastus until the present. It is true that this operation is clearly attested in rabbinic sources and that it had its own name in Greek, Arabic, Aramaic, and Mishnaic Hebrew. Nevertheless, the meaning of the biblical term is not that specific. The parallels to בולס that are closest in etymology and syntax suggest that it refers to the entire process of harvesting sycomore figs, beginning with the gashing.

The etymology of בולס was discovered by Bochart in 1663. It is the participle of a denominative verb, derived from a noun בלס meaning "fig (incl. sycomore fig)." Although Bochart was not able to prove that Hebrew had such a noun, he was able to point to the noun *balas* "fig" in Ethiopic and Arabic. It has always been assumed that this noun is unattested in Hebrew, but this assumption is incorrect. The noun בלס appears in the Mishnah, according to some textual witnesses, as the name of an inferior fig. In one commentary, it is even identified with the sycomore. Moreover, Yemeni Arabic also has a denominative verb derived from the word for "fig": *ballasa* "to pick figs." Its participle *miballis* refers to a person who picks figs from a tree and sells them in the market. Hence, there are no longer any grounds for doubting Bochart's theory.

Bochart's discovery is more significant than he realized. The Hebrew noun בלס and the Yemeni noun *balas* are not cognates, descended from Proto-West-Semitic. The same goes for Hebrew שקמה and Yemeni *súggama* "sycomore." It seems likely that both the Hebrew terms and the Yemeni ones were borrowed from Epigraphic South Arabian. In fact, one of them is attested in Qatabanian, in the phrase *ʾlhw* s^1*qmtm* "sycomore-gods," a phrase that has hitherto been misinterpreted. The ESA term s^1*qmtm* (apparently pronounced something like [šuqamatum]) may, in turn, be derived from the verb **šuqamat* "was made to stand, planted," possibly alluding to a belief that the sycomore was a sacred tree planted by the gods.

These linguistic borrowings appear to have some relevance for the controversies among botanists concerning the origin of the biblical sycomore. They support the traditional view that this tropical tree did not come to Israel spontaneously in the Mesolithic or Neolithic periods but was deliberately introduced in historical times. The linguistic evidence suggests that sycomore figs and/or saplings were imported not from Egypt (as commonly thought) but from Yemen, and that the words for the fig (*bls*) and the tree (*šqmt*) were imported with them. (Yemen is the only place in the world outside of Africa where the fruit of the sycomore produces viable seeds, and it is believed to be the source of other Palestinian trees, such as *Commiphora opobalsamum*, which also grew in royal groves). This must have taken place at some point during the two centuries preceding Solomon's reign, when the first signs of trade with Arabia appear in the archeological record of Israel.

Amos uses the participle בוקר to describe his other occupation. All of the ancient and medieval exegetes took this participle as a denominative derived from the word בקר "cattle." Many modern scholars believe that this interpretation makes 7:14 contradict both 7:15 and 1:1. However, the term נקדים used in 1:1 does not mean "shepherds," as these scholars assume. According to the Targum and some of the Greek translators, it is a synonym of אנשי מקנה and thus may refer to people who deal with בקר or צאן or both. This interpretation is corroborated by the Akkadian cognate. The Neo-Babylonian *nāqidu* was a specialist breeder of בקר (*nāqidu ša lâti*) or צאן (*nāqidu ša ṣēni*) or both (*nāqidu ša ṣēni u lâti*). Amos belonged to the last category. In 7:14, he alludes to his בקר as a sign of self-sufficiency. In 7:15, he alludes to his צאן as a symbol of legitimacy, using a phrase from Nathan's oracle in order to associate himself with David, in opposition to Amaziah and Jeroboam.

In Mesopotamia, even *nāqidu*'s hired by a temple to manage its herds and flocks were businessmen—not temple personnel and certainly not cult personnel. This is even more true of the נקדים from Tekoa. Although they were probably regular suppliers of sacrificial animals for the Temple in Jerusalem, neither they nor their animals belonged to the Temple. They were members of families that owned and managed livestock, living together in a settlement similar to the one in Babylonia known as *Kapri-ša-nāqidāti* "Village of herdsmen" and working together in a kind of collective. They seem to have passed down their business for many generations, judging from the reference to the אדירים of Tekoa in Neh 3:5.

One of the most serious problems faced by stockbreeders is the winter feeding of their animals, especially cattle. To solve this problem, one Mesopotamian *nāqidu ša ṣēni u lâti* leased fields for grazing and raised barley on the side, perhaps for use as fodder. It seems likely that the נקדים from Tekoa leased fields containing sycomore trees, possibly in the Jericho Valley, to feed their animals in the winter. The sycomore is uniquely suited to this purpose, since it is the only tree in the region that bears fruit in the winter and since much of its fruit is unfit for humans but good for cattle. Amos—or his workers—could keep his animals in the shelter of the trees and work on the fruit at the same time. He could sell the good figs and use the inedible ones and the leaves as fodder for his animals. Like the modern Egyptian *gemamzi*, who buys the yearly crop of sycomore fruit in advance and does all the work of gashing and picking, he did not own the trees themselves.

Bibliography

Ancient and Medieval Works

Abarbanel (Abravanel), Isaac. *Don Isaac Abrabanel y su comentario al Libro de Amos.* Ed. G. Ruiz. Madrid: UPCM, 1984.

Athenaeus of Naucratis. *The Deipnosophists.* Trans. C. B. Gulick. 7 vols. Cambridge, Mass.: Harvard University Press, 1927-41.

Baġdādī, ʿAbd al-Laṭīf al-. *Kitāb al-ʾifādah wa-l-ʾiʿtibār.* Damascus: Dār Qutaybah, n.d.

———. *Relation de l'Égypt.* Ed. S. de Sacy. Paris: Imprimerie Impériale, 1810.

Bar Bahlul, Hasan. *Lexicon Syriacum auctore Hassano Bar-Bahlule.* Ed. R. Duval. 3 vols. Paris: e Reipublicae typographaeo, 1888-1901.

Dioscorides Pedanius of Anazarbos. *De Materia Medica.* Ed. M. Wellmann. 3 vols. Berlin: Weidmann, 1906-14.

———. *La 'materia médica' de Dioscórides.* Ed. C. E. Dubler and E. Terés. Tetuán/Barcelona: Emporium, 1953-.

Duran, Profiat. אגרות ר' פריפוט דוראן in ספר מעשה אפֿד. Ed. J. Friedländer and J. Kohn. Vienna: J. Holzwarth, 1865.

Eliezer of Beaugency. *Kommentar zu Ezechiel und den XII kleinen Propheten.* Ed. S. A. Poznański. 3 vols. Warsaw: Mekize Nirdamim, 1909-13.

Farḥi, Estori. כפתור ופרח. Ed. A. M. Luncz. 3rd ed. Jerusalem: A. M. Luncz, 1899.

Fāsī, David ben Abraham al-. *Kitāb Jāmiᶜ al-Alfāẓ of David ben Abraham al-Fāsī*. Ed. S. L. Skoss. 2 vols. New Haven: Yale University Press, 1936-45.

Hai Gaon (?). פירוש הגאונים לספר טהרות. Ed. J. N. Epstein. Jerusalem/Tel-Aviv: Magnes/Devir, 1982.

Ḥimyarī, Našwān bin Saᶜīd al-. *Šams al-ᶜUlūm*. Ed. K. V. Zettersteén. Leiden: E. J. Brill, 1951-53.

Ibn al-Bayṭār, ᶜAbd Allāh ibn Aḥmad. *Traité des simples d'Ibn al-Baïtār de Malaga*. Ed. Mohamed al-Arbi al-Khattabī. N. p.: Dar al-Gharb al-Islami, 1990.

———. *Traité des simples, par Ibn el-Beïthar*. Trans. L. Leclerc. 3 vols. Paris: Impr. nationale, 1877-83.

Ibn al-Kalbī, Hišām. *Kitāb al-aṣnām*. Ed. A. Zeki. Cairo: Dār al-Kutub, 1924.

———. *The Book of Idols*. Trans. N. A. Faris. Princeton: Princeton University Press, 1952.

Ibn ᶜAqnīn, Joseph. התגלות הסודות והופעת המאורות: פירוש שיר השירים Ed. A. S. Halkin. Jerusalem: Mekize Nirdamim, 1964.

Ibn Balᶜam, Judah. Commentary on Minor Prophets. In S. A. Poznański, "The Arabic Commentary of Abu Zakariya Yaḥya (Judah ben Samuel) Ibn Balᶜam on the Twelve Minor Prophets." *JQR* 15 (1924-25) 1-53.

Ibn Danān, Saadia. *Sefer ha-šorašim*. Ed. M. J. Sánchez. Granada: Universidad de Granada, 1996.

Ibn Durayd, Muḥammad. *Kitāb jamharat al-luġah*. Ed. R. M. al-Baᶜlabakkī. 3 vols. Beirut: Dār al-ᶜIlm lil-Malāyīn, 1987-88.

Ibn Ezra, Abraham. שני פירושי ר׳ אברהם אבן עזרא לתרי־עשר. Ed. U. Simon. Ramat-Gan: Bar-Ilan University, 1989-.

Ibn Janāḥ, Jonah. *Kitāb al-ʾuṣūl: The Book of Hebrew Roots*. Oxford: Clarendon Press, 1875.

Ibn Parḥon, Solomon b. Abraham. מחברת הערוך. Ed. S. G. Stern. Pressburg: Typis Antonii Nobilis de Schmid, 1844.

Ibn Quraysh, Judah. ה׳רִסַאלַה׳ של יהודה בן קוּרַיְש. Ed. D. Becker. Tel-Aviv: Tel-Aviv University, 1984.

Ibn Sīnā, Abū ᶜAlī al-Ḥusayn (Avicenna). *Al-Qānūn fi al-ṭibb*. 3 vols. Baghdad: Al-Muthanna Library, n.d.

Isaac b. Melchizedek of Siponto. פירוש הריבמ״ץ לרבנו יצחק ב״ר מלכי צדק מסימפונט למשנה זרעים. Ed. N. Sacks. Jerusalem: Institute for the Complete Israeli Talmud, 1975.

Isaiah b. Mali, di Trani. פירוש נביאים וכתובים. Ed. A. J. Wertheimer. 3 vols. 2nd ed. Jerusalem: Ktab Yad Wasepher, 1978.

Ishodad of Merv. *Commentaire d'Išoʿdad de Merv sur l'Ancien Testament*. Ed. C. Van den Eynde. CSCO 126, 156, 176, 179, 229-. Louvain: L. Durbecq, 1950-.

Jerome, St. *Commentarii in Prophetas Minores*. 2 vols. CC Series Latina 76-76A. Turnholti: Brepols, 1969-70.

———. *S. Eusebii Hieronymi Stridonensis Presbyteri commentariorum in Epistolam ad Titum*. PL 26. Ed. J. P. Migne. Paris: Garnier, 1884.

Maimonides, Moses. משנה עם פירוש רבינו משה בן מימון, מקור ותרגום. Ed. J. Qafiḥ. 6 vols. Jerusalem: Mossad Harav Kook, 1963-68.

———. "Moses Maimonides' Glossary of Drug Names." In *Maimonides' Medical Writings*. Haifa: The Maimonides Research Institute, 1995.

Menaḥem b. Saruq. *Maḥberet*. Ed. A. Sáenz-Badillos. Granada: Universidad de Granada, 1986.

Midrash Deuteronomy Rabba. מדרש דברים רבה. Ed. S. Lieberman. 2nd ed. Jerusalem: Shalem, 1992.

Midrash Genesis Rabba. מדרש בראשית רבא. Ed. J. Theodor and C. Albeck. Jerusalem: Wahrmann, 1965.

Midrash Sifre. ספרי דבי רב. Ed. H. S. Horovitz. Leipzig: G. Fock, 1917.

Mishnah. משנה זרעים עם שינויי נוסחאות מכתבי יד של המשנה. Ed. N. Sacks. 2 vols. Jerusalem: Institute for the Complete Israeli Talmud, 1972-75.

Nathan b. Yeḥiel of Rome. ערוך השלם. Ed. A. Kohut. 8 vols. Vienna: n.p., 1878-92.

Origen. *Origenis Hexaplorum*. Ed. F. Field. Oxford: Clarendon Press, 1875.

Peshitta. *The Old Testament in Syriac According to the Peshiṭta Version*. Leiden: E. J. Brill, 1972–.

Qimḥi, Joseph. ספר הגלוי. Ed. H. J. Mathews. Berlin: Mekize Nirdamim, 1887.

Qumisi, Daniel b. Moses al-. פתרון שנים עשר. Ed. I. D. Markon. Jerusalem: Mekize Nirdamim, 1957.

Rashi. פרשנ־דתא והוא פירוש רש״י על נ״ך. Ed. I. Maarsen. 3 vols. Amsterdam: M. Hertzberger, 1930-.

Rashi, Eliezer of Beaugency, Ibn Ezra, David Qimḥi and Ibn Caspi. *Comentarios hebreos medievales al libro de Amos*. Ed. G. Ruiz González. Madrid: UPCM, 1987.

Saadia Gaon. *Oeuvres complètes de R. Saadia ben Iosef al-Fayyoûmî.* Ed. J. Derenbourg. 5 vols. Paris: E. Leroux, 1893-99.

Sherira Gaon. קטע מפירוש מלים של רב שרירא לסדרי טהרות וזרעים. In תשובות הגאונים מכתבי־יד שבגנזי קמברידג׳. Ed. S. Assaf. Jerusalem: Mekize Nirdamim, 1942. 172-79.

Sirillo, Solomon b. Joseph. מסכת דמאי מן תלמוד ירושלמי . . . לר׳ שלמה בכ״ר יוסף סיריליאו . Ed. P. Shapiro and J. Freimann. Jerusalem: מסורה, 1955.

———. מסכת תרומות מן תלמוד ירושלמי . . . לר׳ שלמה בכ״ר יוסף סיריליאו. Ed. P. Shapiro and J. Freimann. Benei Berak: הרשקוביץ, 1958.

Strabo. *The Geography of Strabo*. Trans. H. L. Jones. 8 vols. Cambridge, Mass.: Harvard University Press, 1949.

Syrohexapla. *Codex Syro-hexaplaris Ambrosianus photolithographice editus*. Ed. A. M. Ceriani. *Monumenta sacra et profana* 7. Milan: Impensis Bibliothecae Ambrosianae, 1874.

Talmud Yerushalmi. *Synopse zum Talmud Yerushalmi*. Ed. P. Schäfer and H.-J. Becker. Tübingen: J. C. B. Mohr, 1991-.

———. תלמוד ירושלמי יוצא לאור על פי כתב יד סקליגר 3 (Or. 4720) שבספריית האוניברסיטה של ליידן עם השלמות ותיקונים. Jerusalem: The Academy of the Hebrew Language, 2001.

Tanḥum b. Joseph ha-Yerushalmi. פירוש תנחום בן יוסף הירושלמי לתרי־עשר. Ed. H. Shy. Jerusalem: Magnes, 1991.

Targum. *The Aramaic Version of Qohelet*. Ed. É. Levine. New York: Sepher-Hermon, 1978.

———. *The Bible in Aramaic*. Ed. A. Sperber. 4 vols. Leiden: E. J. Brill, 1959-73.

———. התרגום השומרוני לתורה. Ed. A. Tal. 3 vols. Tel-Aviv: Tel-Aviv University, 1980-83.

Theodore, Bishop of Mopsuestia. *Theodori Mopsuesteni Commentarius in XII Prophetas*. Wiesbaden: Otto Harrassowitz, 1977.

Theophrastus. *Enquiry into Plants*. Trans. A. Hort. 2 vols. Cambridge, Mass.: Harvard University Press, 1948.

Torah Shelemah. תורה שלמה. Ed. M. Kasher. 42 vols. Jerusalem: n. p., 1927-

Yāqūt ibn ʿAbd Allāh. *Jacut's Geographisches Wörterbuch*. Ed. F. Wüstenfeld. 6 vols. Leipzig: F. A. Brockhaus, 1866-73.

Modern Works

Abramson, S. מפי בעלי לשונות. Jerusalem: Mossad Harav Kook, 1988.

Academy of the Hebrew Language. *Materials for the Dictionary, Series I: 200 B.C.E. — 300 C.E.* Jerusalem: Academy of the Hebrew Language, 1988.

———. מאגרים: CD-ROM 1; Second Century B.C.E. — Mid-Fifth Century C.E. Jerusalem: Academy of the Hebrew Language, 1998.

Akwaʿ, I. al-. *Al-ʾamṯāl al-yamāniyya*. Cairo: Dār al-Maʿārif, 1968.

Allony, N. מחקרי לשון וספרות. 6 vols. Jerusalem: Ben Zvi Institute, 1986-.

Alon, G. *The Jews in their Land in the Talmudic Age*. Trans. Gershon Levi. 2 vols. Jerusalem: Magnes, 1980-84.

Alpin, P. *Plantes d'Egypte*. Trans. R. de Fenoyl. Cairo: Institut français archéologie orientale du Caire, 1980.

Alter, R. "Putting Together Biblical Narrative." Bilgray lecture, University of Arizona, 10 March 1988.

Amar, Z. פירושים ריאליים בפרשנות ימי־הביניים לצומח של ארץ־ישראל. *Sinai* 116 (1995) 86-96.

———. גידולי ארץ־ישראל בימי־הביניים: תיאור ותמורות. Ph. D. diss., Bar-Ilan University, 1996.

———. הגידולים החקלאיים על־פי תבליט לכיש. *Bet Mikra* 159 (1998-99) 350-56.

Andersen, F. I. and D. N. Freedman. *Amos*. AB 24A. New York: Doubleday, 1989.

Anttila, R. *An Introduction to Historical and Comparative Linguistics*. New York: Macmillan, 1972.

Applebaum, S. "Economic Life in Palestine." In *The Jewish People in the First Century: Historical Geography, Political History, Social, Cultural and Religious Life and Institutions*. Ed. S. Safrai and M. Stern. CRINT 1. Assen/Amsterdam: Van Gorcum, 1976. 631-700.

Appleyard, D. L. "Ethiopian Semitic and South Arabian: Towards a Re-examination of a Relationship." *Israel Oriental Studies* 16 (1996) 203-28.

Arieti, J. A. "The Vocabulary of Septuagint Amos." *JBL* 93 (1974) 338-47.

Ashbel, D. הערות לנבואות עמוס. *Bet Mikra* 25 (1965-66) 103-7.

Assaf, S. פירוש ששה סדרי משנה לרבינו נתן אב הישיבה. *Kiryat Sefer* 10 (1934) 381-88, 525-45.

Audo, T. *Sīmtā d-leššānā sūryāyā*. 2 vols. Mossoul: Imprimerie des pères dominicains, 1897.

Auld, A. G. *Amos*. Sheffield: JSOT Press, 1986.

Banitt, M. *Rashi: Interpreter of the Biblical Letter*. Tel-Aviv: Tel Aviv University, 1985.

Bartina, S. "'Hiendo los higos de los sicomoros' (Am 7, 14)." *EstBib* 25 (1966) 349-54.

Baudissin, W. W. *Studien zur semitischen Religionsgeschichte*. 2 vols. Leipzig: F. W. Grunow, 1876-78.

Baum, N. *Arbres et arbustes de l'Egypte ancienne*. Leuven: Departement Oriëntalistiek, 1988.

Baumer, M. *Notes on Trees and Shrubs in Arid and Semi-Arid Regions*. Rome: Food and Agriculture Organization of the United Nations, 1983.

Baur, G. *Der Prophet Amos*. Giessen: J. Ricker, 1847.

Beeston, A. F. L. "On the Correspondence of Hebrew *ś* to ESA s^2." *JSS* 22 (1977) 50-57.

———. *Sabaic Grammar*. JSS Monograph 6. Manchester: JSS, 1984.

———, M. A. Ghul, W. W. Müller, and J. Ryckmans. *Sabaic Dictionary*. JSS Monograph 6. Louvain-la-Neuve: Peeters, 1982.

Behnstedt, P. *Die nordjemenitischen Dialekte*. 2 vols. Wiesbaden: L. Reichert, 1985-92.

Bekele-Tesemma, A. *Useful Trees and Shrubs for Ethiopia*. n.p.: Regional Soil Conservation Unit, Swedish International Development Authority, 1993.

Ben-Ḥayyim, Z. עברית וארמית נוסח שומרון. 5 vols. Jerusalem: Bialik/The Academy of the Hebrew Language, 1957-77.

———. *A Grammar of Samaritan Hebrew*. Winona Lake, Ind.: Eisenbrauns, 2000.

Ben-Yehuda, E. מלון הלשון העברית הישנה והחדשה. 8 vols. New York/London: T. Yoseloff, 1960.

Berg, C. C. "Annotated Check-list of the *Ficus* Species of the African Floristic Region, with Special Reference and a Key to the Taxa of Southern Africa." *Kirkia* 13 (1990) 253-91.

Berg, C. C. and J. T. Wiebes. *African Fig Trees and Fig Wasps*. Amsterdam: North-Holland, 1992.

Bič, M. "Der Prophet Amos—Ein Haepatoskopos." *VT* 1 (1951) 293-96.

———. *Das Buch Amos*. Berlin: Evangelische Verlagsanstalt, 1969.

Bickerman, E. *From Ezra to the Last of the Maccabees*. New York: Schocken, 1962.

Biella, J. C. *Dictionary of Old South Arabic*. HSS 25. Chico, Calif.: Scholars Press, 1982.

Blau, J. "Marginalia Semitica I." *Israel Oriental Studies* 1 (1971) 1-35. Reprinted in J. Blau. *Topics in Hebrew and Semitic Linguistics*. Jerusalem: Magnes, 1998. 185-219.

Blenkinsopp, J. "Did the Second Jerusalemite Temple Possess Land?" *Transeuphratène* 21 (2001) 61-68.

———. בעיות בתורת ההגה והצורות של עברית המקרא: פתרונות קודמים וחדשים. In דברי האקדמיה הלאומית הישראלית למדעים 9 (2001) 1-12.

Blondheim, D. S. "Échos du judéo-hellénisme." *REJ* 78 (1924) 1-14.

Bochart, S. *Hierozoicon*. 2 vols. London: J. Martyn & J. Allestry, 1663.

———. *Hierozoicon*. Ed. E. F. C. Rosenmüller. 3 vols. Leipzig: Weidmann, 1793-96.

Borowski, O. *Agriculture in Iron Age Israel*. Winona Lake, Ind.: Eisenbrauns, 1987.

Breslavy, Y. עמוס—נוקד, בוקר ובולס שקמים. *Bet Mikra* 31 (1966-67) 87-101.

Brockelmann, K. *Lexicon Syriacum*. 2nd ed. Halis Saxonum: M. Niemeyer, 1928.

Bron, F. "Le bilinguisme en Arabie du Sud préislamique." In *Mosaïque de langues, mosaïque culturelle: le bilinguisme dans le Proche-Orient ancien*. Ed. F. Briquel-Chatonnet. Paris: Maisonneuve, 1996. 125-30.

Brown, J. P. *The Lebanon and Phoenicia, The Physical Setting and the Forest*. Beirut: American University of Beirut, 1969.

Brown T. W. and F. G. Walsingham. "The Sycamore Fig in Egypt." *The Journal of Heredity* 8 (1917) 3-12.

Budde, K. "Die Ueberschrift des Buches Amos und des Propheten Heimat." In *Semitic Studies in Memory of Rev. Dr. Alexander Kohut*. Ed. G. A. Kohut. Berlin: S. Calvary, 1897. 106-10.

Busson, F. *Plantes alimentaires de l'ouest africain: étude botanique, biologique et chimique*. Marseille: L'Imprimerie Leconte, 1965.

Cazelles, H. "Mari et l'Ancien Testament." In *La Civilisation de Mari: XVe Rencontre assyriologique international . . . 1966*. Ed. J.-R. Kupper. Paris: Les Belles Lettres, 1967. 73-90.

Collenette, S. *Flowers of Saudi Arabia*. London: Scorpion, 1985.

Condit, I. J. *The Fig*. Waltham, Mass.: Chronica Botanica, 1947.

Craigie, P. C. "Amos the *nōqēd* in the Light of Ugaritic." *SR* 11 (1982) 29-33.

———. *Ugarit and the Old Testament.* Grand Rapids: Eerdmans, 1983.
Cripps, R. S. *A Critical and Exegetical Commentary on the Book of Amos.* 2nd ed. London: SPCK, 1955.
Crum, W. E. *Coptic Dictionary.* Oxford: Oxford University Press, 1939.
Cunchillos, J.-L. and J.-P. Vita. *Concordancia de Palabras Ugaríticas.* Madrid-Zaragoza: Consejo Superior de Investigaciones Científicas, 1995.
Cutler, B., and J. Macdonald. "Identification of the *Naʿar* in the Ugaritic Texts." *UF* 8 (1976) 27-35.
———. "The Unique Ugaritic Text UT 113 and the Question of 'Guilds.'" *UF* 9 (1977) 13-30.
Dalman, G. *Arbeit und Sitte in Palästina.* 7 vols. Gütersloh: C. Bertelsmann, 1928-39.
Danell, G. A. "Var Amos verkligen en nabi?" *SEÅ* 16 (1951) 7-20.
Danin, A. השקמה אינה עץ בר, *Teva Vaaretz* 32 (1990) 28-31.
———. "The Origins of Israel's Sycomores." *Israel Land and Nature* 16 (1990/91) 58-62.
Daumas, M. J. E. *La vie arabe et la société musulmane.* Paris: M. Lévy frères, 1869.
DeGuglielmo, A. "Sacrifice in the Ugaritic Texts." *CBQ* 17 (1955) 76-96.
De Lange, N. R. M. "Some New Fragments of Aquila on Malachi and Job?" *VT* 30 (1980) 291-94.
———. "The Jews of Byzantium and the Greek Bible." In *Rashi 1040-1990: Hommage à Ephraïm E. Urbach.* Ed. G. Sed-Rajna. Paris: Cerf, 1993. 203-10.
———. "La tradition des 'révisions juives' au moyen âge: les fragments hébraïques de la Geniza du Caire." In *"Selon les Septante": Hommage à Marguerite Harl.* Ed. G. Dorival and O. Munnich. Paris: Cerf, 1995. 133-43.
Delcor, M. "Quelques termes relatifs à l'élevage des ovins en hébreu classique et dans les langues sémitiques voisines: étude de lexicographie comparée." In *Atti del secondo congresso internazionale di linguistica camito-semitica.* Ed. P. Fronzaroli. Florence: Università di Firenze, 1978. 105-24.
Dhorme, E. *L'évolution religieuse d'Israël.* Bruxelles: Nouvelle société d'éditions, 1937.

Diebner, B. J. "Berufe und Berufung des Amos (Am 1,1 und 7,14f.)." *Dielheimer Blätter zum Alten Testament und seiner Rezeption in der Alten Kirche* 23 (1986) 97-120.

Dietrich, M. and O. Loretz. "Die ug. Berufsgruppe der *nqdm* und das Amt des *rb nqdm*." *UF* 9 (1977) 336-37.

Dillmann, A. *Lexicon Linguae Aethiopicae*. Leipzig: T. O. Weigel, 1865.

Dixon, D. M. "The Transplantation of Punt Incense Trees in Egypt." *JEA* 55 (1969) 55-65.

Driel, G. van. "Neo-Babylonian Sheep and Goats." *Bulletin on Sumerian Agriculture* 7 (1993) 219-58.

———. "Cattle in the Neo-Babylonian Period." *Bulletin on Sumerian Agriculture* 8 (1995) 215-40.

Driver, S. R. *An Introduction to the Literature of the Old Testament*. New York: Charles Scribner's Sons, 1891.

———. *The Books of Joel and Amos*. Cambridge: Cambridge University Press, 1897.

Easton, M. G. *Illustrated Bible Dictionary*. Grand Rapids, Mich.: Baker Book House, 1978.

Ehrlich, A. B. מקרא כפשוטו. 3 vols. Berlin: M. Poppelauer, 1899-1901.

Eissfeldt, O. "The Prophetic Literature." In *The Old Testament and Modern Study*. Ed. H. H. Rowley. Oxford: Oxford University Press, 1951. 115-61.

———. *The Old Testament: An Introduction*. Oxford: B. Blackwell, 1965.

Elat, M. קשרי כלכלה בין ארצות המקרא בימי בית ראשון. Jerusalem: Bialik, 1977.

Ellenbogen, M. *Foreign Words in the Old Testament: Their Origin and Etymology*. London: Luzac, 1962.

Engnell, I. *Studies in Divine Kingship in the Ancient Near East*. Uppsala: Almqvist & Wiksell, 1943.

———. "Amos." In *Svenskt Bibliskt Uppslagsverk*. Gävle: Skolförlaget, 1948. 60-62.

———. "Profetismens ursprung och uppkomst: Ett gammaltestamentligt grundproblem." *Religion och Bibel: Nathan Söderblom-Sällskapets Årsbok* 8 (1949) 1-18.

Erichsen, W. *Demotisches Glossar*. Copenhagen: E. Munksgaard, 1954.

Erman, A., and H. Grapow. *Wörterbuch der aegyptischen Sprache*. 7 vols. Leipzig: J. C. Hinrichs, 1926-63.

Fales, F. M. *Aramaic Epigraphs on Clay Tablets of the Neo-Assyrian Period*. Rome: Università degli studi "La Sapienza," 1986.

Faulkner, R. O. *The Ancient Egyptian Pyramid Texts*. Oxford: Clarendon Press, 1969.

Feliks, Y. (J.) עולם הצומח המקראי. Ramat-Gan: Massada, 1968.

———. החקלאות בארץ ישראל בימי המקרא המשנה והתלמוד. Jerusalem: R. Mass, 1990.

———. טבע וארץ בתנ״ך. Jerusalem: R. Mass, 1992.

———. עצי־פרי למיניהם: צמחי התנ״ך וחז״ל. Jerusalem: R. Mass, 1994.

Figari, A. *Studii scientifici sull'Egitto e sue adiacenze compresa la penisola dell'Arabia Petrea*. 2 vols. Lucca: G. Giusti, 1865.

Finkelstein, I. "Arabian Trade and Socio-Political Conditions in the Negev in the Twelfth-Eleventh Centuries B.C.E." *JNES* 47 (1988) 241-52.

Finkelstein, J. J. "An Old Babylonian Herding Contract and Genesis 31:38f." *JAOS* 88 (1968) 30-36.

Firmage, E. "Zoology." In *ABD* 6. 1109-67.

Fischer, H. "Another example of the verb *nh* 'shelter.'" *JEA* 64 (1978) 131-32.

Food and Agriculture Organization of the United Nations. *Traditional Food Plants: A Resource Book for Promoting the Exploitation and Consumption of Food Plants in Arid, Semi-arid, and Sub-humid Lands of Eastern Africa*. Rome: Food and Agriculture Organization of the United Nations, 1988.

Forskål, P. *Flora Ægyptiaco-Arabica*. Hauniæ: Ex officina Mölleri, 1775.

Fox, J. T. *Semitic Noun Patterns*. HSS 52. Winona Lake, Ind.: Eisenbrauns, 2003.

Fox, M. Z. (Fox, H.). המשנה בתימן; כתוב־יד מפירוש רב נתן אב הישיבה. *Asufot* 8 (1994) 161-67.

———. תשלום המלקט מפירוש רב נתן אב הישיבה למשנה. In לראש יוסף: מחקרים בחכמת ישראל. Ed. J. Tobi. Jerusalem: Afikim, 1995. 371-86.

Fraenkel, S. *Beiträge zur Erklärung der mehrlautigen Bildungen im Arabischen*. Leiden: E. J. Brill, 1878.

———. *Die aramäischen Fremdwörter im Arabischen*. Leiden: E. J. Brill, 1886.

Freedman, D. N. "Headings in the Books of the Eighth-century Prophets." *AUSS* 25 (1987) 9-26.

Fuhs, H. F. "Amos 1,1: Erwägungen zur Tradition und Redaktion des Amosbuches." In *Bausteine biblischer Theologie: Festgabe für G. Johannes Botterweck*. Ed. H.-J. Fabry. Köln/Bonn: P. Hanstein, 1977. 275.

Gale, R. et al. "Wood." In *Ancient Egyptian Materials and Technology*. Ed. P. T. Nicholson and I. Shaw. Cambridge: Cambridge University Press, 2000. 334-71.

Galil, J. השקמה בתרבות ישראל. *Teva Vaaretz* 8 (1965-66) 306-18, 338-55.

———. "An Ancient Technique for Ripening Sycomore Fruit in East-Mediterranean Countries." *Economic Botany* 22 (1968) 178-90.

———. המוצא של צומח הסאואנות בארץ. *Teva Vaaretz* 14 (1971-72) 139-45.

———. הפיקוס: עץ בר ועץ נוי. Jerusalem: משרד החינוך והתרבות, 1985.

Galil, J. and D. Eisikowitch. "Flowering Cycles and Fruit Types of *Ficus Sycomorus* in Israel." *The New Phytologist* 67 (1968) 745-58.

———. "On the Pollination Ecology of *Ficus Sycomorus* in East Africa." *Ecology* 49 (1968) 259-69.

———. "Further Studies on the Pollination Ecology of *Ficus Sycomorus* L. (Hymenoptera, Chalcidoidea, Agaonidae)." *Tijdschrift voor Entomologie* 112 (1969) 1-13.

Galil, J., M. Stein and A. Horovitz. "On the Origin of the Sycomore Fig (*Ficus sycomorus* L.) in the Middle East." *The Gardens' Bulletin, Singapore* 29 (1976) 191-205.

Galling, K. "Die syrisch-palästinische Küste nach der Beschreibung bei Pseudo-Skylax." *ZDPV* 60 (1937) 66-96.

Gaster, T. H. "Notes on Ras Shamra Texts." *OLZ* 38 (1935) 474-78.

Geiger, A. "Bibliographische Anzeigen." *ZDMG* 12 (1858) 363.

Germer, R. *Flora des pharaonischen Ägypten*. Mainz am Rhein: P. von Zabern, 1985.

Gesenius, W. *Hebräisches and chaldäisches Handwörterbuch über das Alte Testament*. 2nd ed. Leipzig: F. C. W. Vogel, 1823.

———. *Thesaurus philologicus criticus linguae Hebraeae et Chaldaeae Veteris Testamenti*. Leipzig: F. C. W. Vogel, 1835.

Gessel, B. H. L. van. *Onomasticon of the Hittite Pantheon*. Leiden: E. J. Brill, 1998.

Goldmann, F. *La figue en Palestine à l'époque de la mischna*. Paris: Librairie Durlacher, 1911.

Goldstein, J. A. *II Maccabees*. AB 41A. Garden City, N.Y.: Doubleday, 1983.

Goor, A. "The History of the Fig in the Holy Land from Ancient Times to the Present Day." *Economic Botany* 19 (1965) 124-35.

Gordis, R. "Studies in the Relationship of Biblical and Rabbinic Hebrew." In *Louis Ginzberg Jubilee Volume*. New York: American Academy for Jewish Research, 1945. 173-99.

Green, M. W. "Animal Husbandry at Uruk in the Archaic Period." *JNES* 39 (1980) 1-35.

Greenberg, M. *Understanding Exodus*. New York: Behrman House, 1969.

Greenfield, J. C. "Lexicographical Notes I." *HUCA* 29 (1958) 203-28.

———. "Ugaritic Lexicographical Notes." *JCS* 21 (1967) 89-93.

———. "Amurrite, Ugaritic and Canaanite." In *Proceedings of the International Conference on Semitic Studies held on Jerusalem, 19-23 July 1965*. Jerusalem: The Israel Academy of Sciences and Humanities, 1969. 92-101.

Greenspahn, F. E. *Hapax Legomena in Biblical Hebrew: A Study of the Phenomenon and Its Treatment Since Antiquity with Special Reference to Verbal Forms*. SBLDS 74. Chico, Calif.: Scholars Press, 1984.

Grossman, A. ר׳ שמעיה השושני ופירושו לשיר השירים. In ספר היובל לרב מרדכי ברויאר. Ed. M. Bar-Asher. 2 vols. Jerusalem: Academon, 1992. 1. 27-62.

Hadidi, M. N. el- and L. Boulos, *The Street Trees of Egypt*. Cairo: The American University in Cairo Press, 1988.

Haldar A. *Associations of Cult Prophets Among the Ancient Semites*. Uppsala: Almqvist & Wiksell, 1945.

Halpern, B. "The Construction of the Davidic State: An Exercise in Historiography." In *The Origins of the Ancient Israelite States*. Ed. V. Fritz and P. R. Davies. JSOTSup 228. Sheffield: Sheffield Academic Press, 1996. 44-75.

Hammershaimb, E. *The Book of Amos*. Oxford: B. Blackwell, 1970.

Hareuveni, E. גמזיות. *Leš* 11 (1940-41) 39-41.

Hareuveni, N. *Tree and Shrub in Our Biblical Heritage*. Trans. H. Frenkley. Kiriat Ono, Israel: Neot Kedumim, 1984.

Harper, W. R. *A Critical and Exegetical Commentary on Amos and Hosea*. ICC. New York: Scribner, 1905.

Harrington, D. J. and A. J. Saldarini. *Targum Jonathan of the Former Prophets*. Wilmington, Del.: M. Glazier, 1987.

Harrison, R. K. "Sycamore; Sycamore Tree." In *International Standard Bible Encyclopedia*. Ed. G. W. Bromiley. 4 vols. Fully revised ed. Grand Rapids, Mich.: W. B. Eerdmans, 1979. 4. 674.

Hasel, G. F. *Understanding the Book of Amos*. Grand Rapids, Mich.: Baker, 1991.

Haupt, P. "Was Amos a Sheepman?" *JBL* 35 (1916) 280-87.

Hayes, J. H. *Amos, the Eighth-Century Prophet*. Nashville: Abingdon, 1988.

Held, M. "Studies in Comparative Semitic Lexicography." In *Studies in Honor of Benno Landsberger on his Seventy-fifth Birthday April 21, 1965*. AS 16. Chicago: The University of Chicago Press, 1965. 395-406.

Heller, D. and C. C. Heyn. *Conspectus Florae Orientalis: An Annotated Catalogue of the Flora of the Middle East*. Jerusalem: The Israel Academy of Sciences and Humanities, 1994.

Heltzer, M. "Royal Economy in Ancient Ugarit." In *State and Temple Economy in the Ancient Near East, II*. Ed. E. Lipiński. Orientalia Lovaniensia Analecta 6. Leuven: Departement Oriëntalistiek, 1979. 459-96.

———. המשק המלכותי של דוד המלך לעומת המשק המלכותי של אוגרית. *ErIsr* 20 (1989) 175-80.

Henslow, G. "Egyptian Figs." *Nature* 47 (1892) 102.

———. "The Sycomore Fig." *Journal of the Royal Horticultural Society* 27 (1902) 128-31.

Hepper, F. N. "An Ancient Expedition to Transplant Living Trees." *Journal of the Royal Horticultural Society* 92 (1967) 435-38.

———. *Illustrated Encyclopedia of Bible Plants*. Leicester: Inter-Varsity Press, 1992.

———. *Pharaoh's Flowers*. London: HMSO, 1990.

Hitzig, F. *Die zwölf Kleinen Propheten*. 4th ed. Leipzig: S. Hirzel, 1881.

Hoch, J. E. *Semitic Words in Egyptian Texts of the New Kingdom and Third Intermediate Period*. Princeton, NJ: Princeton University Press, 1994.

Hoffmann, G. "Versuche zu Amos." *ZAW* 3 (1883) 87-125.

Höfner, M. and J. M. Solá Solé. *Inschriften aus dem Gebiet zwischen Mārib und dem Ǧōf*. Sammlung Eduard Glaser 2. Vienna: H. Böhlhaus, 1961.

Holladay, J. S., Jr. "The Kingdoms of Israel and Judah: Political and

Economic Centralization in the Iron IIA-B (ca. 1000-750 BCE)." In *The Archaeology of Society in the Holy Land*. Ed. T. E. Levy. London: Leicester University Press, 1998. 368-98.

Hopf, M. "Plant Remains and Early Farming in Jericho." In *The Domestication and Exploitation of Plants and Animals*. Ed. P. J. Ucko and G. W. Dimbleby. Chicago: Aldine, 1969. 355-59.

Hopkins, D.C. *The Highlands of Canaan: Agricultural Life in the Early Iron Age*. Social world of biblical antiquity series 3. Sheffield: Almond, 1985.

Hubaishi, A. al- and K. Müller-Hohenstein. *An Introduction to the Vegetation of Yemen: Ecological Basis, Floristic Composition, Human Influence*. Eschborn: Deutsche Gesellschaft für Technische Zusammenarbeit, 1984.

Hugonot, J.-C. *Le jardin dans l'Egypte ancienne*. Frankfurt am Main: P. Lang, 1989.

———. "Ägyptische Gärten." In *Der Garten von der Antike bis zum Mittelalter*. Ed. M. Carroll-Spillecke. Mainz am Rhein: P. von Zabern, 1992. 9-44.

Humbert, P. "בּוֹלֵס שִׁקְמִים (Amos VII, 14)." *OLZ* 20 (1917) 296-98.

Hurvitz, A. בין לשון ללשון: לתולדות לשון המקרא בימי בית שני. Jerusalem: Bialik, 1972.

ʿInān, Zayd ibn ʿAli. *Al-Lahja al-yamāniyya fī al-nukat wa-l-ʾamṯāl al-ṣanʿāniyya*. n.p.: 1980.

Jamme, A. "Le panthéon sud-arabe préislamique d'après les sources épigraphiques." *Le Muséon* 60 (1947) 57-147.

Japhet, S. *I & II Chronicles*. OTL. London: SCM, 1993.

Jastrow, M. *A Dictionary of the Targumim, the Talmud Babli and Yerushalmi, and the Midrashic Literature*. London: Luzac, 1903.

Jaruzelska, I. *Amos and the Officialdom in the Kingdom of Israel*. Poznań: Wydawnictwo Naukowe Uniwersytetu im. Adama Mickiewicza, 1998.

Jeffers, A. *Magic and Divination in Ancient Palestine and Syria*. Leiden: E. J. Brill, 1996.

Johnstone, T. M. *Jibbāli Lexicon*. Oxford: Oxford University Press, 1981.

Joüon, P. and T. Muraoka. *A Grammar of Biblical Hebrew*. 2 vols. Subsidia Biblica 14. Rome: Pontifical Biblical Institute, 1991.

Kallai, Z. "Tekoa." In אנציקלופדיה מקראית. 9 vols. Jerusalem: Bialik, 1972-88. 8. 924-26.

Kapelrud, A. S. *Central Ideas in Amos*. Oslo: Universitetsforlaget, 1961.

Kasovsky, C. Y. אוצר לשון המשנה. 4 vols. Tel-Aviv: Massadah, 1967.

Kaufman, S. A. *The Akkadian Influences on Aramaic*. AS 19. Chicago/London: University of Chicago Press, 1974.

Kees, H. *Der Götterglaube im alten Ägypten*. Berlin: Akademie-Verlag, 1987.

Keil, C. F. and F. Delitzsch. *The Twelve Minor Prophets*. Trans. J. Martin. Edinburgh: T. & T. Clark, 1868.

Keimer, L. "Eine Bemerkung zu Amos 7,14." *Bib* 8 (1927) 441-44.

———. "An Ancient Egyptian Knife in Modern Egypt." *Ancient Egypt* (1928) 65-66.

———. "Sprachliches und Sachliches zu ελκω 'Frucht der Sykomore.'" *AcOr* 6 (1928) 288-304.

———. "Sur quelques petits fruits en faïence émaillée datant du Moyen Empire." *Bulletin de l'Institut français d'archéologie orientale* 28 (1929) 49-97.

Kelecha, W. M. *A Glossary of Ethiopian Plant Names*. 4th ed. Addis Ababa: n.p., 1987.

Kerdelhué, C., et al. "Comparative Community Ecology Studies on Old World Figs and Fig Wasps." *Ecology* 81 (2000) 2832-49.

King, P. J. *Amos, Hosea, Micah—An Archaeological Commentary*. Philadelphia: The Westminster Press, 1988.

Kislev, M. ההסטוריה של החרוב בארץ. *Halamish* 6 (1988) 20-30.

———. השקמים אשר בתבליטי לכיש — זיהוי בראי דברי חז"ל. In ירושלים וארץ־ישראל: ספר אריה קינדלר. Ed. J. Schwartz, Z. Amar and I. Ziffer. Tel-Aviv: Eretz Israel Museum, 2000. 23-30.

Klapper, R., G. Posner, and M. Friedman. "Amnon and Tamar: A Case Study in Allusions." *Nahalah: Yeshiva University Journal for the Study of Bible* 1 (1999) 23-33.

Kleven, T. "The Cows of Bashan: A Single Metaphor at Amos 4:1-3." *CBQ* 58 (1996) 215-27.

Koemoth, P. *Osiris et les arbres*. Liège: C.I.P.L., 1994.

Kraeling, E. G. "Two Place Names of Hellenistic Palestine." *JNES* 7 (1948) 199-201.

Kraus, F. R. *Staatliche Viehhaltung im altbabylonischen Lande Larsa*. Amsterdam: Noord-Hollandsche U. M., 1966.

Kümmel, H. M. *Familie, Beruf und Amt im spätbabylonischen Uruk*. Berlin: Mann, 1979.

Kuniholm, P. I. "Wood." In *The Oxford Encyclopedia of Archaeology in the Near East*. Ed. E. M. Meyers. 5 vols. New York: Oxford University Press, 1997. 5. 347-49.

Kutscher, E. Y. הלשון והרקע הלשוני של מגילת ישעיהו השלמה ממגילות ים המלח. Jerusalem: Magnes, 1959.

———. לשונן של האיגרות העבריות והארמיות של בר כוסבה ובני דורו. *Leš* 25 (1960-61) 117-33; 26 (1961-62) 7-23.

———. מלים ותולדותיהן. Jerusalem: Kiryath Sefer, 1965.

———. "Mittelhebräisch und jüdisch Aramäisch im neuen Köhler-Baumgartner." In *Hebräische Wortforschung: Festschrift zum 80. Geburtstag von Walter Baumgartner*. VTSup 16. Leiden: E. J. Brill, 1967. 158-75.

———. "Aramaic." In *Current Trends in Linguistics*. Ed. T. A. Sebeok. 14 vols. The Hague: Mouton, 1963-. 6. 354.

Lagarde, P. de. "Ueber die semitischen Namen des Feigenbaums und der Feige." In *Mittheilungen*. 4 vols. Goettingen: Dieterichsche Sortimentsbuchhandlung, 1884-91. 1. 58-75.

Lagrange, M.-J. *Études sur les religions sémitiques*. 2nd ed. Paris: V. Lecoffre, 1905.

Lane, E. W. *Arabic-English Lexicon*. London: Williams and Norgate, 1863-77.

Landberg, C. von. *Glossaire Daṯinois*. 3 vols. Leiden: E. J. Brill, 1920-42.

Landsberger, B. *Materialien zum sumerischen Lexikon, II: Die Serie Ur-e-a* = nâqu. Rome: Pontificium Institutum Biblicum, 1951.

Lanfranchi, G. B. and S. Parpola. *The Correspondence of Sargon II, Part II: Letters from the Northern and Northeastern Provinces*. State Archives of Assyria 5. Helsinki: Helsinki University, 1990.

Le Houérou, H. N. "The Role of Browse in the Sahelian and Sudanian Zones." In *Browse in Africa: The Current State of Knowledge*. Ed. H. N. Le Houérou. Addis Ababa: International Livestock Centre for Africa, 1980. 83-100.

Leslau, W. *Etymological Dictionary of Gurage (Ethiopic)*. 3 vols. Wiesbaden: O. Harrassowitz, 1979.

———. *Comparative Dictionary of Geʿez*. Wiesbaden: O. Harrassowitz, 1987.

Levy, J. *Neuhebräisches und chaldäisches Wörterbuch über die Talmudim und Midraschim.* 4 vols. Leipzig: F. A. Brockhaus, 1876-89.

———. *Chaldäisches Wörterbuch über die Targumim.* 2 vols. Leipzig: Baumgärtner, 1867-68.

Lewy, H. *Die semitischen Fremdwörter im Griechischen.* Berlin: R. Gaertner, 1895.

Lieberman, S. תוספתא כפשוטה: באור ארוך לתוספתא. 10 vols. New York: Jewish Theological Seminary, 1955-88.

Liddell, H. G. and R. Scott. *A Greek-English Lexicon.* Oxford: Clarendon Press, 1996.

Liphschitz, N. "*Ceratonia Siliqua* in Israel: An Ancient Element or a Newcomer." *Israel Journal of Botany* 36 (1987) 191-97.

Liphschitz, N. and G. Biger. השקמה בישראל בעת העתיקה לפי ממצאים בוטניים מחפירות. *Hassadeh* 72 (1992) 771-72.

Liver, Y. פרשת מחצית השקל. In ספר היובל ליחזקאל קויפמן. Ed. M. Haran. Jerusalem: Magnes, 1961. 54-67.

Loret, V. "Les livres III et IV (animaux et végétaux) de la *Scala Magna* de Schams-ar-Riâsah (1re partie)." *Annales du service des Antiquités de l'Egypte* 1 (1899) 48-63.

Loretz, O. "Die Berufung des Propheten Amos (7,14-15)." *UF* 6 (1974) 487-88.

Löw, I. *Aramäische Pflanzennamen.* Leipzig: W. Engelmann, 1881.

———. *Die Flora der Juden.* 4 vols. Vienna and Leipzig: R. Lowit, 1924-34.

Lucas, A. *Ancient Egyptian Materials and Industries.* 4th ed. London: Histories and Mysteries of Man, 1989.

Ludolf, H. *Ad suam Historiam æthiopicam antehac editam Commentarius.* Frankfurt am Main: J. D. Zunner, 1691.

Macdonald, M. C. A. and L. Nehmé. "Al-ʿUzzā." In *Encyclopaedia of Islam.* Leiden: E. J. Brill, 1954-. 10. 967-68.

Mankowski, P. V. *Akkadian Loanwords in Biblical Hebrew.* HSS 47. Winona Lake, Ind.: Eisenbrauns, 2000.

Margoliouth, G. *Catalogue of the Hebrew and Samaritan Manuscripts in the British Museum.* 4 vols. London: The British Museum, Dept. of Oriental Printed Books and Manuscripts, 1899-1935.

Masterman, E. W. G. "Sycomore, Tree." In *International Standard Bible Encyclopaedia.* Ed. J. Orr. 5 vols. Chicago: Howard-Severance, 1915. 5. 2877.

Maydell, H.-J. von. *Trees and Shrubs of the Sahel.* Eschborn: Deutsche Gesellschaft für Technische Zusammenarbeit, 1986.

Melamed, E. Z. "The Breakup of Stereotyped Phrases as an Artistic Device in Biblical Poetry." *Scripta Hierosolymitana* 8 (1961) 115-53.

Mendelsohn, I. "Guilds in Ancient Palestine." *BASOR* 80 (December, 1940) 17-21.

The Merck Index: An Encyclopedia of Chemicals, Drugs, and Biologicals. Rahway, N.J.: Merck and Co., 1989.

Mettinger, T. N. D. *Solomonic State Officials.* ConBOT 5. Lund: CWK Gleerup, 1971.

Millard, A. R. "Assyrian Royal Names in Biblical Hebrew." *JSS* 21 (1976) 1-14.

Miller, A. G. and T. A. Cope. *Flora of the Arabian Peninsula and Socotra.* Edinburgh: Edinburgh University Press, 1996-.

Moftah, R. *Die heiligen Bäume im Alten Ägypten.* Ph. D. diss., Georg-August-Universität zu Göttingen, 1959.

———. "Die uralte Sykomore und andere Erscheinungen der Hathor." *Zeitschrift für ägyptische Sprache und Altertumskunde* 92 (1966) 40-47.

Moldenke, H. N. and A. L. Moldenke. *Plants of the Bible.* Waltham, Mass.: Chronica Botanica, 1952.

Montgomery, J. A. "Notes on Amos." *JBL* 23 (1904) 94-96.

———. "The New Sources of Knowledge." In *Record and Revelation.* Ed. H. W. Robinson. Oxford: Clarendon Press, 1938. 1-27.

Moorey, P. R. S. *Ancient Mesopotamian Materials and Industries.* Oxford: Clarendon Press, 1994.

Moreshet, M. לקסיקון הפועַל שנתחדש בלשון התנאים. Ramat-Gan: Bar-Ilan University Press, 1980.

Morgenstern, J. "Amos Studies I." *HUCA* 11 (1936) 19-140.

Morrison, M. A. "Evidence for Herdsmen and Animal Husbandry in the Nuzi Documents." In *Studies on the Civilization and Culture of Nuzi and the Hurrians.* Ed. M. A. Morrison and D. I. Owen. Winona Lake, Ind.: Eisenbrauns, 1981-. 1. 257-96.

Müller, C. "Holz und Holzverarbeitung." In *Lexikon der Ägyptologie.* Ed. W. Helck and E. Otto. 7 vols. Wiesbaden: O. Harrassowitz, 1975-92. 2. 1264-69.

Murray, M. A. "Fruits, Vegetables, Pulses and Condiments." In *Ancient Egyptian Materials and Technology.* Ed. P. T. Nicholson and I. Shaw. Cambridge: University Press, 2000. 609-55.

Murtonen, A. "The Prophet Amos—A Hepatoscoper?" *VT* 2 (1952) 170-71.

Muschler, R. *A Manual Flora of Egypt*. Berlin: R. Friedlaender & Sohn, 1912.

Myers, J. M. *I Chronicles*. AB 12. 1st ed. Garden City, N.Y.: Doubleday, 1965.

Na'aman, N. "Sources and Composition in the History of David." In *The Origins of the Ancient Israelite States*. Ed. V. Fritz and P. R. Davies. JSOTSup 228. Sheffield: Sheffield Academic Press, 1996. 170-86.

Naval Intelligence Division. *Iraq and the Persian Gulf*. Geographical Handbook Series, B.R. 524, 1944.

Nöldeke, T. Review of F. Delitzsch, *Prolegomena eines neuen hebräisch-aramäischen Wörterbuchs zum Alten Testament*. *ZDMG* 40 (1886) 718-43.

Noth, M. *Die israelitischen Personennamen im Rahmen der gemeinsemitischen Namengebung*. Stuttgart: W. Kohlhammer, 1928.

Ochsenschlager, E.L. "Sheep: Ethnoarchaeology at Al-Hiba." *Bulletin on Sumerian Agriculture* 7 (1993) 33-42.

Olmo Lete, G. del and J. Sanmartín. *Diccionario de la lengua ugarítica*. Aula Orientalis-Supplementa 8. 2 vols. Sabadell, Barcelona: Editorial AUSA, 1996-2000.

Pardee, D. "The Baʿlu Myth." In *The Context of Scripture*. Ed. W. W. Hallo and K. L. Younger. 3 vols. Leiden: E. J. Brill, 1997-2002. 1. 241-74.

Pastor, J. *Land and Economy in Ancient Palestine*. London: Routledge, 1997.

Paul, S. M. *Amos*. Hermeneia. Minneapolis: Fortress, 1991.

Payne Smith, R. *Thesaurus Syriacus*. 2 vols. Oxford: Clarendon Press, 1879-1901.

Piamenta, M. *Dictionary of Post-Classical Yemeni Arabic*. Leiden: E. J. Brill, 1990.

Post, G. E. *Flora of Syria, Palestine and Sinai*. 2nd ed. by J. E. Dinsmore. Beirut: American Press, 1932-33.

Postgate, J. N. "Some Old Babylonian Shepherds and Their Flocks." *JSS* 20 (1975) 1-21.

Pulcini, T. *Exegesis as Polemical Discourse: Ibn Ḥazm on Jewish and Christian Scriptures*. Atlanta, Ga.: Scholars Press, 1998.

Qimron, E. *The Hebrew of the Dead Sea Scrolls*. HSS 29. Atlanta, Ga.: Scholars Press, 1986.

Rabin, C. *Ancient West-Arabian*. London: Taylor's Foreign Press, 1951.

———. "The Nature and Origin of the Šafʿel in Hebrew and Aramaic." *ErIsr* 9 (1969) 148-58.

Rahmani, L. Y. *A Catalogue of Jewish Ossuaries in the Collections of the State of Israel*. Jerusalem: Israel Antiquities Authority, 1994.

Rahmer, M. "Die hebräischen Traditionen in den Werken des Hieronymus, Zweiter Theil: Die Commentarien zu den XII kleinen Propheten, III. Amos." *MGWJ* 42 (1898) 1-16, 97-107.

Rainey, A. F. מבנה החברה באוגרית. Jerusalem: Bialik, 1967.

Redding, R. W. "Subsistence Security as a Selective Pressure Favoring Increasing Cultural Complexity." *Bulletin on Sumerian Agriculture* 7 (1993) 77-98.

Reider, J. *Prolegomena to a Greek-Hebrew and Hebrew-Greek Index to Aquila*. Philadelphia: n.p., 1916. Reprinted in *Studies in the Septuagint: Origins, Recensions and Interpretations*. Ed. Sidney Jellicoe. New York: Ktav, 1974.

Renfrew, J. M. *Palaeoethnobotany: The Prehistoric Food Plants of the Near East and Europe*. New York: Columbia University Press, 1973.

Reynier, L. "Méthode de caprification usitée sur le figuier sycomore." In *Mémoires sur l'Egypte publiés dans les campagnes du général Bonaparte*. 4 vols. Paris: P. Didot L'ainé, 1800-1803. 3. 184-89.

Ricks, S. D. *Lexicon of Inscriptional Qatabanian*. Rome: Editrice Pontificio Istituto Biblico, 1989.

Robert, P. de. *Le berger d'Israël*. Neuchâtel: Delachaux et Niestlé, 1968.

Rosenbaum, S. N. *Amos of Israel: A New Interpretation*. Macon, Ga.: Mercer University Press, 1990.

Rosenfeld, A. *The Authorised Selichot for the Whole Year*. London: I. Labworth, 1962.

Rosenthal, E. S. בירורי מלים וחילופי נוסח. In מחקרי תלמוד: קובץ מחקרים מוקדש לזכרו של פרופ׳ אליעזר שמשון רוזנטל . . . . Ed. M. Bar-Asher and D. Rosenthal. Jerusalem: Magnes, 1993. 13-46.

Rudolph, W. *Joel—Amos—Obadja—Jona*. KAT 13/2. Gütersloh: G. Mohn, 1971.

Ryckmans, G. "Inscriptions sud-arabes, cinquième série." *Le Muséon* 52 (1939) 51-112.

Sachs, S. "A Tree Drooping with its Ancient Burden of Faith." In *The New York Times*, Dec. 26, 2001, A4.

Sáenz-Badillos, A. *A History of the Hebrew Language*. Trans. J. Elwolde. Cambridge: Cambridge University Press, 1993.

Safrai, S. "The Temple and the Divine Service." In *The World History of the Jewish People: The Herodian Period*. Ed. M. Avi-Yonah. New Brunswick, N.J.: Rutgers University Press, 1975. 282-337.

———. "The Temple." In *The Jewish People in the First Century: Historical Geography, Political History, Social, Cultural and Religious Life and Institutions*. Ed. S. Safrai and M. Stern. CRINT 1. Assen/Amsterdam: Van Gorcum, 1976. 865-907.

Salonen, A. Review of W. von Soden, *Akkadisches Handwörterbuch*. *AfO* 23 (1970) 95-97.

Sanders, J. A. *The Psalms Scroll of Qumrân Cave 11*. DJD IV. Oxford: Clarendon Press, 1965.

San Nicolò, M. "Materialien zur Viehwirtschaft in den neubabylonischen Tempeln. I." *Or* NS 17 (1948) 273-93.

———. "Materialien zur Viehwirtschaft in den neubabylonischen Tempeln. II." *Or* NS 18 (1949) 288-306.

———. "Materialien zur Viehwirtschaft in den neubabylonischen Tempeln. III." *Or* NS 20 (1951) 129-50.

———. "Materialien zur Viehwirtschaft in den neubabylonischen Tempeln. IV." *Or* NS 23 (1954) 351-82.

———. "Materialien zur Viehwirtschaft in den neubabylonischen Tempeln. V." *Or* NS 25 (1956) 24-38.

Schäfer-Lichtenberger, C. "Sociological and Biblical Views of the Early State." In *The Origins of the Ancient Israelite States*. Ed. V. Fritz and P. R. Davies. JSOTSup 228. Sheffield: Sheffield Academic Press, 1996. 78-105.

Schlesinger, A. כתבי עקיבא שליזנגר: מחקרים במקרא ובלשונו. Jerusalem: Israel Society for Biblical Research, 1962.

Schmidt, H. *Der Prophet Amos*. Tübingen: J. C. B. Mohr, 1917.

Schult, H. "Amos 7_{15a} und die Legitimation des Aussenseiters." In *Probleme biblischer Theologie: Gerhard von Rad zum 70. Geburtstag. . . .* Ed. H. W. Wolff. Munich: C. Kaiser, 1971. 462-78.

Schulthess, F. *Lexicon Syropalaestinum*. Berlin: G. Reimer, 1903.

Schwartz, G. M. "Pastoral Nomadism in Ancient Western Asia." In *Civilizations of the Ancient Near East*. Ed. J. M. Sasson. 4 vols. New York: Scribner, 1995. 1. 249-58.

Schweinfurth, G. "Sitzungs-Bericht vom 15. October 1889." *Sitzungs-Berichte der Gesellschaft naturforschender Freunde zu Berlin* (1889) 157-59.

———. "Sammlung arabisch-æthiopischer Pflanzen." *Bulletin de l'Herbier Boissier* 4 (1896) Appendix II, 91-203.

———. "Über die Bedeutung der 'Kulturgeschichte.'" *Botanische Jahrbücher* 45 (1910), Beiblatt 103. 28-38.

Segert, S. "Zur Bedeutung des Wortes nōqēd." In *Hebräische Wortforschung: Festschrift zum 80. Geburtstag von Walter Baumgartner*. VTSup 16. Leiden: E. J. Brill, 1967. 279-83.

———. "The Ugaritic *nqdm* After Twenty Years. A Note on the Function of Ugaritic *nqdm*." *UF* 19 (1987) 409-10.

Selwi, I. al-. *Jemenitische Wörter in den Werken von al-Hamdānī und Našwān und ihre Parallelen in den semitischen Sprachen*. Berlin: D. Reimer, 1987.

Shapira, Y. גילן של שקמים עתיקות. *Teva Vaaretz* 9 (1966-67) 28-29.

Sickenberger, E. *Contributions à la Flore d'Égypte. Mémoires présentés à l'Institut Égyptien* 4/2. Cairo: 1901.

Silver, M. *Prophets and Markets: The Political Economy of Ancient Israel*. Boston: Kluwer-Nijhoff, 1983.

Sjöberg, E. "De förexiliska profeternas förkunnelse." *SEÅ* 14 (1949) 7-42.

Smith, G. V. *Amos: A Commentary*. Grand Rapids, Michigan: Zondervan, 1989.

Soggin, J. A. *The Prophet Amos*. London: SCM, 1987.

Sokoloff, M. *A Dictionary of Jewish Palestinian Aramaic of the Byzantine Period*. Ramat-Gan, Israel: Bar Ilan University Press, 1990.

Solms-Laubach, H. Grafen zu. *Die Herkunft, Domestication und Verbreitung des gewöhnlichen Feigenbaums (Ficus Carica L.)*. Abhandlungen der Königlichen Gesellschaft der Wissenschaften zu Göttingen 28. Göttingen: Dieterich, 1882.

Sommer, B. D. *A Prophet Reads Scripture: Allusion in Isaiah 40-66*. Contraversions. Stanford: Stanford University Press, 1998.

Speier, S. "Bemerkungen zu Amos." *VT* 3 (1953) 305-10.

Spiegelberg, W. "Drei demotische Schreiben aus der Korrespondenz des Pherendates, des Satrapen Darius' I., mit den Chnum-Priestern von Elephantine." *SPAW* 30 (1928) 604-22.

Stager, L. E. "Forging an Identity: The Emergence of Ancient Israel." In *The Oxford History of the Biblical World*. Ed. M. D. Coogan. New York/Oxford: Oxford University Press, 1998. 123-75.

Stark, F. "Some Pre-Islamic Inscriptions on the Frankincense Route in Southern Arabia." *JRAS* 1939 479-98.

Steiner, R. C. *The Case for Fricative-Laterals in Proto-Semitic*. AOS 59. New Haven: American Oriental Society, 1977.

———. "A Colloquialism in Jer 5:13 from the Ancestor of Mishnaic Hebrew." *JSS* 37 (1992) 11-26.

———. "*Ketiv-Ḳere* or Polyphony: The שׁ-שׂ Distinction According to the Masoretes, the Rabbis, Jerome, Qirqīsanī, and Hai Gaon." In *Studies in Hebrew and Jewish Languages Presented to Shelomo Morag*. Ed. M. Bar-Asher. Jerusalem: Bialik, 1996. *151-*179.

———. "Ancient Hebrew." In *The Semitic Languages*. Ed. R. Hetzron. London: Routledge, 1997. 145-73.

———. "*Albounout* 'Frankincense' and *Alsounalph* 'Oxtongue': Phoenician-Punic Botanical Terms from an Egyptian Papyrus and a Byzantine Codex." *Or* 70 (2001) 97-103.

Stępień, M. *Animal Husbandry in the Ancient Near East: A Prosopographic Study of Third-Millennium Umma*. Bethesda, Md.: CDL, 1996.

Stern, M. *Greek and Latin Authors on Jews and Judaism*. 3 vols. Jerusalem: The Israel Academy of Sciences and Humanities, 1974-84.

Stoebe, H. J. "Der Prophet Amos und sein bürgerlicher Beruf." *Wort und Dienst: Jahrbuch der Theologischen Schule Bethel* 51 (1957) 160-81.

Stol, M. "Old Babylonian Cattle." *Bulletin on Sumerian Agriculture* 8 (1995) 173-213.

Stowe, C. E. "Sycamore." In *Dr. William Smith's Dictionary of the Bible*. Ed. H. B. Hackett. 4 vols. New York: Hurd and Houghton, 1870. 4. 3130-31.

Streck, M. P. *Das amurritische Onomastikon der altbabylonischen Zeit*. AOAT 271. Münster: Ugarit-Verlag, 2000-.

Täckholm, V. *Faraos blomster*. Stockholm: Generalstabens Litografiska Anstalt, 1969.

Tal, A. לשון התרגום לנביאים ראשונים ומעמדה בכלל ניבי הארמית. Tel-Aviv: Tel-Aviv University, 1975.

———. *A Dictionary of Samaritan Aramaic*. 2 vols. Leiden: E. J. Brill, 2000.

Talshir, D. מעמדה של העברית המקראית המאוחרת בין לשון המקרא ללשון חכמים. *Language Studies* (מחקרים בלשון) 2-3 (1987) 161-72.

Tarragon, J.-M. de. *Le Culte à Ugarit*. CahRB 19. Paris: J. Gabalda, 1980.

Tawil, H. "Late Hebrew-Aramaic ספר, Neo-Babylonian *sirpu/sirapu*: A Lexicographical Note IV." *Bet Mikra* 154-55 (1997-98) 339-44.

Tawil, H. M. al-. *Early Arab Icons: Literary and Archaeological Evidence for the Cult of Religious Images in Pre-Islamic Arabia*. Ph. D. diss., University of Iowa, 1993.

Teixidor, J. *The Pagan God*. Princeton, N.J.: Princeton University Press, 1977.

———. *The Pantheon of Palmyra*. Leiden: E. J. Brill, 1979.

Thompson, H. O. "Carmel, Mount." In *ABD* 1. 874-75.

Thompson, R. C. *A Dictionary of Assyrian Botany*. London: The British Academy, 1949.

Tigay, J. H. "'Heavy of Mouth' and 'Heavy of Tongue': On Moses' Speech Difficulty." *BASOR* 231 (1978) 57-67.

Tirosh-Becker, O. "Linguistic Study of a Rabbinic Quotation Embedded in a Karaite Commentary on Exodus." In *Studies in Mishnaic Hebrew*. Ed. M. Bar-Asher. Jerusalem: Magnes, 1998. 380-407.

Tov, E. "The Septuagint." In *Mikra*. Ed. M. J. Mulder. Assen: Van Gorcum, 1988. 161-88.

Townsend, C. C. and E. Guest. *Flora of Iraq*. Vol. 4. Baghdad: Ministry of Agriculture, 1980.

Treu, U. "Amos VII 14, Schenute und der Physiologos." *NovT* 10 (1968) 234-40.

Tristram, H. B. *Natural History of the Bible: Being a Review of the Physical Geography, Geology, and Meteorology of the Holy Land; With a Description of Every Animal and Plant Mentioned in Holy Scripture*. 2nd ed. London: Society for Promoting Christian Knowledge, 1868.

Tur-Sinai, N. H. (X-VII) בשולי המלון של אליעזר בן־יהודה *Leš* 13 (1944-45) 95-119.

———. הלשון והספר: כרך הלשון. 2nd ed. Jerusalem: Bialik, 1954.

———. פשוטו של מקרא. 4 vols. Jerusalem: Kiryath Sefer, 1962-68.

Ullmann, M. *Wörterbuch der klassischen arabischen Sprache.* Wiesbaden: O. Harrassowitz, 1970-.
Urie, D. M. L. "Officials of the Cult at Ugarit." *PEQ* 80 (1948) 42-47.
Van De Mieroop, M. "Sheep and Goat Herding According to the Old Babylonian Texts from Ur." *Bulletin on Sumerian Agriculture* 7 (1993) 161-82.
Van der Burgt, J. M. M. *Dictionnaire Français-Kirundi.* Bois-le-Duc, Holland: Société "L'Illustration Catholique," 1903.
Van Wyk, P. *Trees of the Kruger National Park.* 2 vols. Cape Town: Purnell, 1972-74.
Vaux, R. de. *Ancient Israel.* New York: McGraw-Hill, 1961.
Virolleaud, C. "Fragment nouveau du poème de Môt et Aleyn-Baal." *Syria* 15 (1934) 226-43.
Waetzoldt, H. "Hirt. A. Philologisch (neusumerisch)." In *Reallexikon der Assyriologie.* Ed. E. Ebeling, B. Meissner, et al. Berlin/Leipzig: W. de Gruyter, 1928-, 4. 421-25.
Warnekros, H. E. "Historia naturalis sycomori ex veterum botanicorum monumentis et itinerariis conscripta." *Repertorium für Biblische und Morgenländische Litteratur* 11 (1782) 224-71; 12 (1783) 81-104.
Watts, J. D. W. *Vision and Prophecy in Amos.* Expanded Anniversary ed. Macon, Ga.: Mercer University Press, 1997.
Weiblen, G. D. *Phylogeny and Ecology of Dioecious Fig Pollination.* Ph.D. diss., Harvard University, 1999.
Weinberg, J. *The Citizen-Temple Community.* JSOTSup 151. Sheffield: Sheffield Academic Press, 1992.
Weippert, H. "Amos: Seine Bilder und ihr Milieu." In *Beiträge zur prophetischen Bildsprache in Israel und Assyrien.* Ed. H. Weippert, K. Seybold, and M. Weippert. Freiburg: Universitätsverlag, 1985.
Weisberg, D. B. *Guild Structure and Political Allegiance in Early Achaemenid Mesopotamia.* Yale Near Eastern researches 1. New Haven: Yale University Press, 1967.
Weiss, M. ספר עמוס. 2 vols. Jerusalem: Magnes, 1992.
Wellhausen, J. *Reste arabischen Heidentums.* Berlin: W. de Gruyter, 1961.
Werth, E. "Die 'wilde' Feige im östlichen Mittelmeergebiet und die Herkunft der Feigenkultur." *Berichte der Deutschen Botanischen Gesellschaft* 50 (1932) 539-57.

Western, A. C. "The Ecological Interpretation of Ancient Charcoals from Jericho." *Levant* 3 (1971) 31-40.

Wilkinson, A. *The Garden in Ancient Egypt*. London: Rubicon, 1998.

Willoughby, B. E. "Amos, Book of." In *ABD* 1. 203-12.

Wilson, R. R. *Prophecy and Society in Ancient Israel*. Philadelphia: Fortress, 1980.

Wiseman, D. J. "Mesopotamian Gardens." *Anatolian Studies* 33 (1983) 137-44.

Wolff, H. W. *Joel and Amos*. Trans. W. Janzen, et al. Ed. S. D. McBride, Jr. Hermeneia. Philadelphia: Fortress, 1977.

Wright, J. "Did Amos Inspect Livers?" *AusBR* 23 (1975) 3-11.

Wright, T. J. "Amos and the 'Sycomore Fig.'" *VT* 26 (1976) 362-68.

Wright, W. *A Grammar of the Arabic Language*. 2 vols. 3rd ed. Cambridge: University Press, 1896-98.

Würthwein, E. "Amos-Studien." *ZAW* 62 (1950) 10-52.

Yeivin, I. מסורת הלשון העברית המשתקפת בניקוד הבבלי. 2 vols. Jerusalem: The Academy of the Hebrew Language, 1985.

Zadok, R. *Geographical Names According to New- and Late-Babylonian Texts*. Wiesbaden: L. Reichert, 1985.

———. "Zur Geographie Babyloniens während des sargonidischen, chaldäischen, achämenidischen und hellenistischen Zeitalters." *WO* 16 (1985) 19-79.

Zalcman, L. "Piercing the Darkness at *Bôqēr* (Amos VII 14)." *VT* 30 (1980) 252-55.

Zimmern, H. *Akkadische Fremdwörter als Beweis für babylonischen Kultureinfluss*. Leipzig: J. C. Hinrichs, 1915.

Ziv, Y. "בוקר ובולס־שקמים"—בתקוע? *Bet Mikra* 92 (1982-83) 49-53.

Zohary, D. "Fig." In *Evolution of Crop Plants*. Ed. J. Smartt and N. W. Simmonds. 2nd ed. New York: Wiley, 1995. 366-70.

——— and M. Hopf. *Domestication of Plants in the Old World*. 3d ed. Oxford: Clarendon Press, 2000.

Zohary, M. *Plant Life of Palestine*. New York: Ronald Press, 1962.

———. *Plants of the Bible*. Cambridge: Cambridge University Press, 1982.

———. "שקמים." In אנציקלופדיה מקראית. 9 vols. Jerusalem: Bialik, 1972-88. 8. 257-58.

Zuntz, L. "Un testo ittita di scongiuri." *Atti del Reale Istituto Veneto di scienze lettere ed arti* 96/2 (1936-37) 477-546.

Index of Subjects

Abarbanel, 5, 46, 71, 77 n. 59, 91, 101 n.36, 105 n. 57, 106 n. 63, 116
Abba Saul, 102 n. 40
ʿAbd al-Laṭīf al-Baġdādī, 16 n. 60, 38 n. 29
ʾAbū Ḥanīfa al-Dīnawar, 38 n. 29
Abū Rāʾiṭa (bishop of Takrit), 26
acacia, as sacred tree, 57
accents, Masoretic, 103
Acer pseudoplatanus, 4
Acre, wheat-stampers of, 96
Adam and Eve, irresistible fruit eaten by, 38 n. 26
Africa, East
  as source of sycomore in Egypt, 49
  sycomore leaves/figs as food for livestock in, 108
Africa, South
  ancient sycomore in, 105 n. 55
  sycomore leaves/figs as food for livestock in, 108
Africa, West, problem of seasonal forage in, 115
Aglibol, sacred cypress at temple of, 57
Ain Feshkha, sycomores at, 110
Akkadian, 3
  cognates of mishnaic botanical terms in, 41
  mušku in, 29, 30 n. 130
  nāqidu in, 73-87, 101 n. 34, 122
Alalakh, *šanannu* at, 86
Alexandria, word for "mulberries" at, 18
allusion, marker of, 93-94
almog, 62, 63
Ālu-ša-nāqidāti (City of herdsmen), 96
Amharic, *bls* in, 53
anvil, sycomore, 27, 28, 46
Aquila, 5, 18, 20-24, 26, 66, 71, 109 n. 80
Arabia. *See* Arabian Peninsula
Arabia, South
  and *almog,* 63
  importation of trees from, 62
Arabian, Epigraphic South
  meaning of s^1qmtm in, 55-58
  and origin of *bls* and *šqmt,* 58-59, 63-65, 121
Arabian, Modern South, 45 n. 65, 53
Arabian Peninsula
  distribution of *balas* in, 53
  distribution of *Ceratosolen* wasp in, 61 n. 65
  distribution of *Ficus* in, 59 n. 53
  and origin of words for "sycomore," 29 n. 128
  trade routes from, 63

Arabic
balas in, 32, 42, 45, 59 n. 54, 59 n. 56, 121
cognate of נוקד in, 72, 73, 81
cognates of mishnaic botanical terms in, 41
order of attributive modifiers in, 90 n. 119
jummayz in, 29, 34
and objection to בוקר as denominative, 67
rare word for "wild fig" in, 13 n. 50
saqima in, 55
sawqam in, 29
tīn in, 38
word for "gashing sycomore figs" in, 120
word for "lisping" in, 25-26
words for "Egyptian willow" in, 54 n. 30
Arabic, Algerian, saqūm in, 58 n. 50
Arabic, Egyptian
gemamzi(a) in, 33
word for "gashing sycomore figs" in, 16 n. 60
Arabic, Yemeni, 3
balas(ah) in, 36 n. 22, 43, 45 n. 66, 53, 55, 59 n. 56, 59 n. 57, 121
jafn in, 55 n. 33
khanas in, 36-37, 39
miballis in, 33, 43, 102 n. 41, 121
soqam in, 53-55
Aramaic
and Akkadian mušku, 29
cognates of mishnaic botanical terms in, 41
Late, and objection to בוקר as denominative, 67 n. 6, 69
meaning of בקרא in, 69
and origin of συκάμινος, 20
targumic, 20, 71
term for "garden of the gods" in, 57
translation of בולס, in Bar Bahlul's dictionary, 11
use of, by Aquila, 22, 23
word for "livestock" derived from Iranian in, 71-72
word for "sycomore branches/figs" in, 34 n. 12
words for "gashing sycomore figs" in, 12, 120
Aramaic, Christian Palestinian
and origin of συκάμινος, 20
שוקמא in, 20, 29, 65
and šqmt, 53
word for "scratch" in, 12
Aramaic, Jewish, rare word for "wild fig" in, 13 n. 50
Argobba, bls in, 53
Aristotle, ψελλός and τραυλός defined by, 25 n. 101
Asaph, compared with Baal-hanan, 31
Asher, tribe of, and location of Tekoa, 101 n. 36
Asir, 61 n. 65
assimilation, in Assyrian word for "sycomore," 29 n. 127
Assur, and cultivation of foreign trees, 52
Assurnasirpal II, interest in trees, 52
Assyria, kings of, and Solomon, 52 n. 22
Assyrian, pronunciation of mušku in, 29
Assyrians, and importation of sycomore beams, 30
asyndeton/syndeton, 89
Athenaeus
and κνίζω, 10
and συκάμινα, 18
attributive modifiers, order of, 89, 90 n. 119
Avestan, 71
Avicenna, 17 n. 63

Baal cycle, נוקד and, 80
Baal-hanan, as overseer of sycomores, 30-31
Babylonia
description of, in Strabo, 30 n. 132
herdsmen in, 78-79, 84-87, 96, 122

Babylonian, Old, and נוקד, 79
Babylonian Talmud. *See* Talmud, Babylonian
Bantus, and sycomore leaves/fruit as food for livestock, 108
Bar Ali, 15 n. 57
Bar Bahlul, 8, 11-12, 13 n. 50, 38 n. 28
Bar Kokhba, sibilants in letters of, 22
Bar Serošwai, 38 n. 28
barren tree, שקמה as, 102 n. 40
bats, fruit, 51
beam(s), construction
  cedar, 28
  sycomore, 24, 27-31, 46, 102, 112, 120
bedouin, 55, 70
booth, of David, 93
branchlets, 34, 35 n. 14
broken plural. *See* plural, broken
Bronze Age
  and aromatics trade, 3, 63
  fig pips from sites of, 50
  as time of introduction of sycomore into Israel, 49
Burundi, cult of sycomore in, 58 n. 49
Byzantium, Jews of, 24

Cairo
  hooked blade for gashing figs in, 11 n. 39
  Virgin's Tree in, 56 n. 44
camel, dromedary, and aromatics trade, 63
Carmel, Mount
  idolatrous tree at, 57
  and Sycaminopolis, 19, 109
carob (tree)
  branchlets of, 35 n. 14
  idolatrous, 57
  imported from South Arabia, 62
  paired with sycomore, 39, 62
  shade from, 106
catholicos, Nestorian, and Hai Gaon, 15 n. 57
causative
  and etymology of *šqmt,* 64-65
  in Qatabanian, 64
cedar, 28, 63
Chalcolithic Period, fig pips from, 50
cheese, made with latex, 41-42
collectives, 96-97, 122
Commiphora opobalsamum, 62, 121
conversational implicature, 91
cooperatives. *See* collectives
Coptic
  Arabic dictionary of, 16 n. 60
  word for "sycomore fig" in, 60
Corbonas, 98
corporations. *See* collectives
cows of Bashan, 117
Crete, Cyprian fig in, 11
cult personnel, herdsmen as, 81-87, 99
cypress, as sacred tree, 57
Cyprus
  "cathedral fig tree" in, 105 n. 55
  instrument used to gouge sycomore fig in, 11
  sycomores in, 7, 120

Daʾūd al-Anṭākī, and oiling of figs, 15 n. 55
David
  Amos associated with, 92-94, 122
  and plans for capital, 31
  and sycomore groves/plantations, 30-31, 48, 104
Dead Sea, and Amos' sycomores, 105, 110
Deir el-Baḥri, sycomore tree at, 56
Demotic (Egyptian), words for "sycomore (fig)" in, 60
denominative(s)
  Aramaic, and בוקר, 68
  ballasa as, 33
  בולס as, 32-33, 35, 42-43, 120
  בוקר as, 32-33, 66-68, 121-122
  Greek, 32
  מגמז as, 33-34
  referring to professions, 69

denominative(s) (*cont.*)
 תעולל as, 46
determinatives, hieratic, and gashed sycomore fig, 10
Dhofar (Oman), 61 n. 65
 absence of *bls* in, 53
 sycomore in, 61 n. 65
Dioscorides
 Arabic translation of, 12 n. 47, 17
 and ἐπικνίζω, 10
 and meaning of συκάμινον, 18
drupelets, 35
Duran, Profiat, 101 n. 36

Eanna temple (Uruk), hersmen of, 74, 78-79, 80, 84-87, 97, 99, 109-10
Ebabbar temple (Sippar), 100
Egypt
 Greek verb in descriptions of sycomore in, 9
 instrument for gashing sycomore figs in, 10-11
 propagation of sycomore in, 51 n. 20, 61
 pyramid texts in, 56
 as source of sycomore in Israel, 49, 59, 61, 121
 study of sycomores in, 7
 and συκάμινος, 18-20
 sycomore leaves and fruit as fodder in, 108
 sycomores for roof timbers in, 31
 temple real estate in, 98 n. 16
 town of Nht in, 57 n. 46
Egypt, Upper, sycomores in, 51 n. 20
Egyptian
 determinatives in, 10
 nbs in, 40 n. 39
 words for "sycomore (fig)" in, 60
Eliezer of Beaugency, 6, 106
Elijah, 105 n. 57
Elisha, 91, 92, 105
enzyme, found in latex, 41-42 n. 49
Epigraphic South Arabian. *See* Arabian, Epigraphic South
Ethiopia
 figs as food for livestock in, 108
 study of sycomores in, 7
 sycomore as sacred tree in, 58
Ethiopian
 balas in, 32, 52-53, 55, 59, 121
 Sabean influence on, 55 n. 33
ethylene, and sycomore figs, 9

Famagusta, "cathedral fig tree" in, 105 n. 55
Fara texts, Sumerian, 74, 75
Fāsī, David al-, 6, 14, 16, 66 n. 3, 76 n. 58
fattener, 118 n. 122
feeding, supplemental, 113
ficin. *See* enzyme, found in latex.
Ficus capreæfolia, 53 n. 24
Ficus carica. See fig, Carian
Ficus cordata, 53 n. 26
Ficus gnaphalocarpa, 3 n. 6, 115
Ficus ingens, 53 n. 26
Ficus johannis, 53 n. 26
Ficus palmata, 53 n. 23
Ficus salicifolia, 54 n. 30
Ficus sycomorus gnaphalocarpa, 3 n. 6, 61 n. 65
Ficus sycomorus sycomorus, 3 n. 6
Ficus trachphylla Fenzl., 3 n. 6
fig
 ass (Eselsfeige), 3
 Carian, 15 n. 55, 35 n. 15, 40, 50, 53 n. 23
 cluster, 3
 Cyprian, 11, 39 n. 34
 Egyptian, 3, 39 n. 34
 fig-mulberry, 3, 19 n. 76
 Lesbian, 40 n. 44
 mulberry, 3
 pharaoh's, 3
 wasps, 8, 9 n. 28, 61
 wild, 3, 13, 38 nn. 27-30
 wild, words for in Jibbāli, 53 n. 25
fig tree, cathedral, 105 n. 55
folk etymology, 19 n. 76
food tree, שקמה as, 102 n. 40

formula, stereotyped, 92, 94
fossils, sycomore, 50

Gedera, as location of David's sycomores, 30 n. 133
Geez, 32, 53, 58 n. 50, 59 n. 57. *See also* Ethiopian
gemamzi(a), 33, 44, 102, 111, 122
Genesis Rabbah, 38 n. 26
genitive construction, 68-69, 90 n. 119
Geonim, Babylonian, and word for "sycomore branches/figs," 34
Gideon, 91
Gilgamesh, shepherd's hut in, 89
gods, irrigation, 55
Greek
- influence on pronunciation of Hebrew from, 22, 26
- meaning and etymology of συκάμινος in, 17-20
- meaning of κνίζω in, 8-10
- among Palestinian Jews, 22
- term for "garden of the gods" in, 57
- translation of בולס in, 8
- translations of נוקד in, 71, 122
- verbs from names of figs in, 32, 43
- words for "scratching" in, 8-10, 12
- ψελλός in, 25

Greeks, and encounter with sycomore, 20
groves, royal, 30-31, 48, 104, 121
guilds. *See* collectives
Guinean zone, and sycomore leaves/fruit as food for livestock, 108
Gurage, *bls* in, 53

Ḥaḍramawt, 54
Hai Gaon, 5, 6, 15, 16 n. 60
al-Hāmī, 54
Hammurabi, Code of, 75, 76 n. 54
harmony, vowel, 37 n. 23
Hatshepsut, and importation of trees, 51
Hebrew, Ephraimite, and sound [š], 27 n. 112
Hebrew, medieval, and word for "sycomore figs," 38
Hebrew, Mishnaic
- ancient botanical and agricultural terms in, 41-42
- בלס in, 36-40, 59 n. 55, 121
- גמזיות in, 29, 34
- descendant of colloquial idiom, 41 n. 45
- metathesis in, 41 n. 44
- מגמז in, 33-35
- and objection to בוקר as denominative, 67
- צמל in, 42
- word for "gashing sycomore figs" in, 12, 120
- word for "gathering olives" in, 47
- words for "sycomore figs" in, 29, 38, 40

Hebrew, Samaritan, and sound [š], 27 n. 112
Hebron, 89, 97, 99
Hejaz, 61 n. 65
hepatoscopist, נוקד and, 83
Herodian period, and sheep in Moab, 77 n. 60
Herodion, 70
Hiram, and *almog*, 63
Ibn al-Kalbī, Hišām, 57 n. 46
Hittite, cedar-gods in, 56
horticulture, sycomore, 3, 10, 17, 120
hypercorrection, *sawqam* as, 54 n. 29

Ibn ᶜAqnīn, Joseph, 15 n. 57
Ibn Balᶜam, Judah, 5, 15, 17
Ibn Durayd, Muḥammad, 54
Ibn Ezra, Abraham, 5, 6, 16-17, 66 n. 3
Ibn Janāḥ, Jonah, 6, 15 n. 57, 66, 72-73, 90, 106
Ibn Parḥon, Solomon, 6, 106
Ibn Quraysh, Judah, 72, 76 n. 58
Ibn Wāfid, ᶜAbd al-Raḥmān, 17 n. 63
Idrīsī, al-, 97 n. 15

inflorescence. *See syconium.*
Iqišā son of Nannā-ereš, 74, 110
Iraq
  sycomore not found in, 16 n. 60, 30
  transhumance in, 113
Iron Age
  integration of animal husbandry and agriculture in, 100
  remains of *Ficus sycomorus* from, 50
Isaac b. Melchizedek of Siponto, 37, 38
Isaiah of Trani, 6
Ishodad of Merv, 13 n. 50, 15 n. 57, 76 n. 56
Isis, 56
Iṣṭaḫrī, al-, 97 n. 15

Jacob, and Laban, 79 n. 70
Jericho and Jericho Valley
  bilingual (Greek-Aramaic) inscription from, 22
  royal groves of *Commiphora opobalsamum* near, 62
  seasonal migration of animals to, 89, 112, 115, 122
  sycomores in, 50, 102 n. 40, 105, 111
Jerome, 5, 6, 8, 24-26, 38 n. 30, 66, 76-77 n. 59, 101, 117. *See also* Vulgate
Jerusalem, 16 n. 60, 88, 89, 97, 99, 115
Jerusalem Temple, 82, 98 n. 16
Jibbāli, words for "wild figs" in, 45 n. 65, 53 n. 25
Jordan Valley, 105 n. 57, 111, 114, 115
Joseph b. Nisan of Shaveh-kiriathaim, 34 n. 11
Josephus
  and royal groves near Jericho, 62
  and פסילוס, 26
  and temple flocks, 98
Joshua b. Qorḥah, 57
juniper, 63

Kalah, and cultivation of foreign trees, 52
Kapru-ša-nāqidāti (Village of herdsmen), 96-97, 122
Kenya, sycomores in, 7
Koine, Egyptian branch of, in LXX, 18 n. 70
Kruger National Park, 105 n. 55, 108 n. 76

Laban, and Jacob, 79 n. 70
Lachish, sycomores depicted in Assyrian reliefs of, 28, 46
Lamb Chamber, of Temple, 98
laryngeals, Greek influence on pronunciation of, 22
latex, 42 n. 50
Latin, *ficus* in, 38
Lebanon, 58 n. 52
  and *almog,* 62, 63
Levi b. Yefet, 5, 14, 16
Leviticus Rabbah, 24 n. 96
lisp, of Amos, 24-25
Al-Luḥayya, 54

magister pecoris, 79 n. 72
Maimonides, 13-14 n. 50, 38 n. 29, 39, 54 n. 30, 102 n. 40
Malakbel, sacred cypress at temple of, 57
Mandaic, and בוקר, 68 n. 9
manuring, of fields, 112 n. 92, 113
maquis bush, Mediterranean, 51
marker, of allusion, 93-94
Masada, sycomore beams at, 28
Masoretic
  accents, 103
  text, 12 n. 45
mass/count nouns, 55
Mecca, sacred trees near, 57
Mekhilta de-R. Simeon bar Yoḥai, 16 n. 59, 33 n. 11
Menaḥem b. Saruq, 6
Mentuhotpe, sycomore tree at temple of, 56
Mesha, as נוקד, 77, 85

Mesopotamia
herdsmen employed by temples in, 74, 78, 82, 84-87, 97, 99-100, 122
and history of word for "sycomore," 29 n. 128
roof-beams in, 101 n. 40
sycomores in, 30 n. 131
timber cutting in, 60 n. 62
See also Akkadian, Babylonia, Iraq, etc.
metathesis, 29 n. 128, 40 n. 39, 41 n. 44
Mishnah, 12, 13 n. 49, 14, 15 n. 55, 19, 27 n. 113, 28 n. 115, 30, 36, 37, 38, 39, 40, 41, 42, 67, 98, 102 n. 40, 106, 112
Moab
herdsmen from, 89
king of, 77
source of rams for Herod's temple, 77 n. 60, 99
Modern South Arabian. *See* Arabian, Modern South
Moses
and צאן, 91-92
speech impediment of, 25-26
and thornbush, 57
mulberry, Greek words for, 17-20
myrrh trees, importation of, 51

Nanna-Ningal temple (Ur), 99-100
Našwān bin Saʿīd al-Ḥimyarī, 53 n. 23
Nathan. *See* oracle, of Nathan
Nathan Av ha-Yeshivah, 13-14 n. 50, 36, 38, 39
Nathan b. Yeḥiel of Rome, 12 n. 49, 37
Natufian (Mesolothic) Period, as time of introduction of sycomore into Israel, 49, 50, 121
Neith (goddess), 56
Neolithic Period
fig pips from sites of, 50
as time of introduction of sycomore into Israel, 49, 121
Nephthys (goddess), 56
Nimrud. *See* Kalah.
Nut (goddess), 56
Nut/Hathor (goddess), 56
Nuzi
hereditary trades at, 95
pasturing of flocks at, 88-89

olive trees, 29, 30-31, 48 n. 1
Oman, *Ficus* in, 58
Onqelos. *See* Targum Onqelos
Ophir, and *almog*, 63
oracle, of Nathan, 93, 94, 122

Palestine
Arabic word for "gashing sycomore figs" in, 16 n. 60
carob in, 62
gashing of sycomore figs in, 14-15, 33-34 n. 11
Greek encounter with sycomore in, 20
sacred sycomore in, 56-57
toponyms containing a word for "sycomore" in, 19 n. 74
Palestinian Talmud. *See* Talmud, Palestinian
palm tree, 30 n. 132
Palmyra, sacred cypress at, 57
panicles. *See* branchlets.
passive, and etymology of *šqmt*, 65
patriarchs, and צאן, 91-92
Peninsula, Arabian. *See* Arabian Peninsula
Peshiṭta, 5, 21, 46-47, 58, 87
Pesiqta de-Rab Kahana, 24 n. 96
Philo, 71 n. 31
Platanus occidentalis, 3-4
Pliny, and royal groves near Jericho, 62
plural, broken, 33 n. 6
Predynastic period (Egypt), 51 n. 20
propagation, clonal (vegetative), 61-62
Proto-Semitic, and *š*-causative, 64
pseudo-Gershom, 102 n. 40
pseudo-Scylax, 19
Punt, and importation of myrrh trees, 51

pyramid texts, 56

Qara, Joseph, 5, 27
Qataban, sacred sycomore in, 55, 57-58
Qatabanian
  s¹qmtm in, 55, 65, 121
  š-causative in, 64
Qimḥi, David, 5, 23, 46, 101 n. 36, 106 n. 36, 107
Qimḥi, Joseph, 5
Queen of Sheba, opobalsamon and, 62
Qumisi, Daniel al-, 6, 76 n. 58
Qumran, and sibilants, 22

Rashi, 5, 23-24, 26-27, 31, 102 n. 40, 120
reduplication, 17
rennet, 42 n. 50
rent, for pasturage, 112, 113, 115
repatterning, 65
Resh Laqish, 107 n. 64

Saadia Gaon, 25-26
Saadia Ibn Danān, 90 n. 123
Sabaic, *qwm* in, 64
Sahel, and sycomore leaves/fruit as food for livestock, 108
Salix aegyptiaca, 54 n. 30
Samson of Sens, 13 n. 49
Sargon II, and interest in trees, 52
Saudi Arabia, *Ficus* in, 58
Saul, 91
seeds, of sycomore, 61
segolate, 37 n. 23, 63 n. 78
Semitic, Proto-West, and origin of *bls* and *šqmt,* 52, 59, 121
Sepphoris, gristmakers of, 96
Septuagint (LXX)
  and Hebrew sibilants, 21
  interpretation of בולס in, 5, 8-11, 15-16, 43-45, 47
  κτηνοτρόφος in, 71
  translation of חפש in, 20
  use of Egyptian branch of koine in, 18 n. 70
  use of κνίζω in, 9-10
Sharon, cattle in, 30 n. 133, 48 n. 1, 99
Sheba, Solomon's ties with, 63
shekels, new, 98
Shemaiah, pupil and amanuensis of Rashi, 24
Shephelah
  livestock in, 103
  sycomores in, 30-31, 48, 101-102
  trade routes to, 63
Sherira Gaon, 6, 13, 15 n. 57, 16 n. 60, 106-8
Shiḥr, al-, 54
sibilants, interchange of, 21-23
Sifra, 101 n. 40
Sifre Zuṭa, 98 n. 18
silviculture, sycomore, 3, 27-28, 120
Silwan, 104 n. 51
Sippar, Ebabbar temple of, 100
Sirillo, Solomon b. Joseph, 13 n. 49, 37, 39
Soddo, *bls* in, 53
Solomon
  Arabia trade in and before time of, 62-63, 121
  importing of trees and wood by, 52 n. 22, 62-63
South Africa. *See* Africa, South
Spanish, Old, word for "Egyptian willow" in, 54 n. 30
speech impediment, of Amos, 24-26
Strabo
  on royal groves of *Commiphora opobalsamum* near Jericho, 62
  on scarcity of timber in Babylonia, 30 n. 132
  on Sycaminopolis, 19, 109
  on sycomore, 18, 38 n. 32
Sudan, sycomore fruits from, 51
Sudanian zone, and sycomore leaves/fruit as food for livestock, 108
Sumerian, na-gada in, 74, 75, 79
superordinate term, 80
superscriptions, of prophetic books, 89 n. 117, 90 n. 119
Suqām, and sacred trees, 57

sycamine, Egyptian, 11, 39 n. 34
Sycaminopolis, 19, 57 n. 46, 109
sycomore-gods, 56, 65
sycomore groves. *See* groves, royal
syconium (*syconia*), 34-35
Symmachus, 5, 18, 66, 71, 76, 109 n. 80
syndeton/asyndeton, 89
syntax
of 2 Chr 26:10, 103
of Amos 1:1, 87-90
of Amos 5:22, 119
of Amos 7:14, 45-46, 68-69
Syria, sycomore foreign to, 58
Syriac
בקרא in, 68
חרט in, 12, 15 n. 57
נקדא in, 74 n. 44
שקמא in, 53, 58
תאנֿא in, 38
תוֿבא in, 13 n. 50, 15 n. 57
Syrohexapla, 12, 15-16
Syropalestinian version, 11-12, 15

Ta'izz, 54
Talmud, Babylonian, 19 n. 75, 27 n. 113, 28, 66, 97, 102 n. 40
Talmud, Palestinian, 14, 25, 34, 37, 38 n. 26, 38 n. 32, 67, 96, 98, 102 n. 40
Tamīmī, 16 n. 60
Tammuz, Seventeenth of, 34 n. 11
Tanḥum Yerushalmi, 6
Targum, 20, 21, 122
Targum, Samaritan, 12 n. 48
Targum Jonathan, 5, 20, 66, 71-72, 76, 101
Targum Onqelos, 20, 72
Tekoa, 67, 70, 82, 87-90, 95, 97, 101-102, 104, 113, 115
Temple mount, coins found on, 99
temple personnel. *See* cult personnel, herdsmen as
Tertiary Period, as time of introduction of sycomore into Israel, 50
Theodoret of Cyrrhus, 5, 8
Theodotion, 18, 21, 66, 109 n. 80
Theophrastus
and carob/sycomore pairing, 39
on the gashing of sycomore figs, 8-11, 17 n. 63, 47, 120
and meaning of συκάμινος, 18
on royal groves of *Commiphora opobalsamum* near Jericho, 62
sycomore seeds and, 61 n. 64
use of ἐπικνίζω by, 10
use of ἐπιτέμνω by, 11
thornbush, and Moses, 57
Tiberias, fishermen of, 96
Tiglath-Pileser I, and cultivation of foreign trees, 51-52
Tigre, *bls* in, 53
Tigrinya, *bls* in, 53
Tosefta, 28, 38 n. 32, 39, 40, 41, 56, 77 n. 60, 96, 98 n. 20, 99, 102 n. 40
trades, hereditary, 95
traitors, class, 118
transhumance, 89, 112-3
treasurers, Temple 99
trees, sacred, 57-58
Tyre, cedar and juniper from, 63

Ugarit
Akkadian texts from, 77
professional groups at, 96
Ugaritic
nqdm in, 77-78, 80-81, 83-86, 96, 104
ṣml in, 42
Ur
hereditary trades at, 95
Nanna-Ningal temple of, 99
Uruk. *See* Eanna temple (Uruk), herdsmen of
Uzziah, 103

virgin sycomore, 46
Virgin's Tree, 56 n. 44
voicing, in Assyrian, 29 n. 127
Vulgate, 21, 87

Wadi Bayḥān, *s¹qmtm* in inscription from, 55

Wady Kelt, sycomores by, 105
wasps, 8-9, 61
willow, Egyptian, 54 n. 30

Yāqūt ibn ʿAbd Allāh, 57 n. 46
Yefet b. Eli, 6
Yemen
  as source of sycomore in Israel, 61-63, 121
  study of sycomores in, 7
  sycomore as sacred tree in, 57-58
  wild sycomores in, 61-62
Yemeni(te). *See* Arabic, Yemeni
Yeshuʿah b. Yehudah, 5, 16, 33-34 n. 11

Zizyphus spina-Christi, 49, 60 n. 60, 65

The Catholic Biblical Quarterly Monograph Series (CBQMS)

1. Patrick W. Skehan, *Studies in Israelite Poetry and Wisdom* (CBQMS 1) $9.00 ($7.20 for CBA members) ISBN 0-915170-00-0 (LC 77-153511)
2. Aloysius M. Ambrozic, *The Hidden Kingdom: A Redactional-Critical Study of the References to the Kingdom of God in Mark's Gospel* (CBQMS 2) $9.00 ($7.20 for CBA members) ISBN 0-915170-01-9 (LC 72-89100)
3. Joseph Jensen, O.S.B., *The Use of tôrâ by Isaiah: His Debate with the Wisdom Tradition* (CBQMS 3) $3.00 ($2.40 for CBA members) ISBN 0-915170-02-7 (LC 73-83134)
4. George W. Coats, *From Canaan to Egypt: Structural and Theological Context for the Joseph Story* (CBQMS 4) $4.00 ($3.20 for CBA members) ISBN 0-915170-03-5 (LC 75-11382)
5. O. Lamar Cope, *Matthew: A Scribe Trained for the Kingdom of Heaven* (CBQMS 5) $4.50 ($3.60 for CBA members) ISBN 0-915170-04-3 (LC 75-36778)
6. Madeleine Boucher, *The Mysterious Parable: A Literary Study* (CBQMS 6) $2.50 ($2.00 for CBA members) ISBN 0-915170-05-1 (LC 76-51260)
7. Jay Braverman, Jerome's Commentary on Daniel: A Study of Comparative Jewish and Christian Interpretations of the Hebrew Bible (CBQMS 7) $4.00 ($3.20 for CBA members) ISBN 0-915170-06-X (LC 78-55726)
8. Maurya P. Horgan, *Pesharim: Qumran Interpretations of Biblical Books* (CBQMS 8) $6.00 ($4.80 for CBA members) ISBN 0-915170-07-8 (LC 78-12910)
9. Harold W. Attridge and Robert A. Oden, Jr., *Philo of Byblos,* The Phoenician History (CBQMS 9) $3.50 ($2.80 for CBA members) ISBN 0-915170-08-6 (LC 80-25781)
10. Paul J. Kobelski, *Melchizedek and Melchireša*ᶜ (CBQMS 10) $4.50 ($3.60 for CBA members) ISBN 0-915170-09-4 (LC 80-28379)
11. Homer Heater, *A Septuagint Translation Technique in the Book of Job* (CBQMS 11) $4.00 ($3.20 for CBA members) ISBN 0-915170-10-8 (LC 81-10085)
12. Robert Doran, *Temple Propaganda: The Purpose and Character of 2 Maccabees* (CBQMS 12) $4.50 ($3.60 for CBA members) ISBN 0-915170-11-6 (LC 81-10084)

13. James Thompson, *The Beginnings of Christian Philosophy: The Epistle to the Hebrews* (CBQMS 13) $5.50 ($4.50 for CBA members) ISBN 0-915170-12-4 (LC 81-12295)
14. Thomas H. Tobin, S.J., *The Creation of Man: Philo and the History of Interpretation* (CBQMS 14) $6.00 ($4.80 for CBA members) ISBN 0-915170-13-2 (LC 82-19891)
15. Carolyn Osiek, *Rich and Poor in the Shepherd of Hermes* (CBQMS 15) $6.00 ($4.80 for CBA members) ISBN 0-915170--14-0 (LC 83-7385)
16. James C. VanderKam, *Enoch and the Growth of an Apocalyptic Tradition* (CBQMS 16) $6.50 ($5.20 for CBA members) ISBN 0-915170-15-9 (LC 83-10134)
17. Antony F. Campbell, S.J., *Of Prophets and Kings: A Late Ninth-Century Document (1 Samuel 1-2 Kings 10)* (CBQMS 17) $7.50 ($6.00 for CBA members) ISBN 0-915170-16-7 (LC 85-12791)
18. John C. Endres, S.J., *Biblical Interpretation in the Book of Jubilees* (CBQMS 18) $8.50 ($6.80 for CBA members) ISBN 0-915170-17-5 (LC 86-6845)
19. Sharon Pace Jeansonne, *The Old Greek Translation of Daniel* 7-12 (CBQMS 19) $5.00 ($4.00 for CBA members) ISBN 0-915170-18-3 (LC 87-15865)
20. Lloyd M. Barré, *The Rhetoric of Political Persuasion: The Narrative Artistry and Political Intentions of* 2 *Kings* 9 -11 (CBQMS 20) $5.00 ($4.00 for CBA members) ISBN 0-915170-19-1 (LC 87-15878)
21. John J. Clabeaux, *A Lost Edition of the Letters of Paul: A Reassessment of the Text of the Pauline Corpus Attested by Marcion* (CBQMS 21) $8.50 ($6.80 for CBA members) ISBN 0-915170-20-5 (LC 88-28511)
22. Craig Koester, *The Dwelling of God: The Tabernacle in the Old Testament, Intertestamental Jewish Literature, and the New Testament* (CBQMS 22) $9.00 ($7.20 for CBA members) ISBN 0-915170-21-3 (LC 89-9853)
23. William Michael Soll, *Psalm* 119*: Matrix, Form, and Setting* (CBQMS 23) $9.00 ($7.20 for CBA members) ISBN 0-915170-22-1 (LC 90-27610)
24. Richard J. Clifford and John J. Collins (eds.), *Creation in the Biblical Traditions* (CBQMS 24) $7.00 ($5.60 for CBA members) ISBN 0-915170-23-X (LC 92-20268)
25. John E. Course, *Speech and Response: A Rhetorical Analysis of the Introductions to the Speeches of the Book of Job, Chaps.* 4 – 24 (CBQMS 25) $8.50 ($6.80 for CBA members) ISBN 0-915170-24-8 (LC 94-26566)

26. Richard J. Clifford, *Creation Accounts in the Ancient Near East and in the Bible* (CBQMS 26) $9.00 ($7.20 for CBA members) ISBN 0-915170-25-6 (LC 94-26565)
27. John Paul Heil, *Blood and Water: The Death and Resurrection of Jesus in John* 18 – 21 (CBQMS 27) $9.00 ($7.20 for CBA members) ISBN 0-915170-26-4 (LC 95-10479)
28. John Kaltner, *The Use of Arabic in Biblical Hebrew Lexicography* (CBQMS 28) $7.50 ($6.00 for CBA members) ISBN 0-915170-27-2 (LC 95-45182)
29. Michael L. Barré, S.S., *Wisdom, You Are My Sister: Studies in Honor of Roland E. Murphy, O.Carm., on the Occasion of His Eightieth Birthday* (CBQMS 29) $13.00 ($10.40 for CBA members) ISBN 0-915170-28-0 (LC 97-16060)
30. Warren Carter and John Paul Heil, *Matthew's Parables: Audience-Oriented Perspectives* (CBQMS 30) $10.00 ($8.00 for CBA members) ISBN 0-915170-29-9 (LC 97-44677)
31. David S. Williams, *The Structure of 1 Maccabees* (CBQMS 31) $7.00 ($5.60 for CBA members) ISBN 0-915170-30-2
32. Lawrence Boadt and Mark S. Smith (eds.), *Imagery and Imagination in Biblical Literature: Essays in Honor of Aloysius Fitzgerald, F.S.C.* (CBQMS 32) $9.00 ($7.20 for CBA members) ISBN 0-915170-31-0 (LC 2001003305)
33. Stephan K. Davis, *The Antithesis of the Ages: Paul's Reconfiguration of Torah* (CBQMS 33) $11.00 ($8.80 for CBA members) ISBN 0-915170-32-9 (LC 2001007936)
34. Aloysius Fitzgerald, F.S.C., *The Lord of the East Wind* (CBQMS 34) $12.00 ($9.60 for CBA members) ISBN 0-915170-33-7 (LC 2002007068)
35. William L. Moran, *The Most Magic Word: Essays on Babylonian and Biblical Literature* (CBQMS 35) $11.50 ($9.20 for CBA members) ISBN 0-915170-34-5 (LC 2002010486)
36. Richard C. Steiner, *Stockmen from Tekoa, Sycomores from Sheba: A Study of Amos' Occupations* (CBQMS 36) ISBN 0-915170-35-3 (LC 2003019378)

Order from:

The Catholic Biblical Association of America
The Catholic University of America
Washington, D.C. 20064